ORIGIN AND DEVELOPMENT

LIFE

ORIGIN AND DEVELOPMENT

GÖSTA EHRENSVÄRD

THE UNIVERSITY OF CHICAGO PRESS

Original edition: *Liv: Ursprung och utformning.* © Gösta Ehrensvärd 1960
Published by Bokförlaget Aldus/Bonniers, Stockholm, 1960. Second edition 1961
Illustrations by Gösta Ehrensvärd, Anders Henschen, and J. W. Sullivan

Library of Congress Catalog Card Number: 62-19620

THE UNIVERSITY OF CHICAGO PRESS, CHICAGO & LONDON
The University of Toronto Press, Toronto 5, Canada

English translation © 1962 by The University of Chicago
All rights reserved. Published 1962. Composed and printed by
THE UNIVERSITY OF CHICAGO PRESS
Chicago, Illinois, U.S.A.

PREFACE

It is obvious that the contents of a book as small as this cannot pretend in any way to fulfil the somewhat ambitious title. Everyone who has worked in any branch of natural science knows only too well how difficult it is to master even superficially the data and methods outside his own special field. Every attempt to analyze the extensive, complicated problems connected with the development of life on our earth, now and in the earliest phases, must therefore be regarded as a very precarious undertaking, for confrontation with data from the most divergent fields of research invites sweeping generalizations of questionable value. Even the most unpretentious demands for completeness and exactness in a comprehensive treatment of the concept of life and its origin would result in a handbook of innumerable volumes, and its contents, presented in technical jargon, would seem dry reading even for the interested scientist. A condensation of all this in a brief presentation of the complex problem in everyday language can never be satisfactorily realized. What can be accomplished will always be marked by a bizarre combination of superficiality and ponderosity.

Further comments are of course superfluous. What is presented here as an exposition of certain questions concerning life-as-a-whole and as chemical activity in cells and organisms must be regarded as an attempt to give an idea of the chemical logic underlying different manifestations of life. This treatment of the subject does not pretend to be more than a series of miscellaneous reflections on certain biochemical problems connected with what is popularly called the origin of life. This has been dealt with somewhat unconventionally, but it may give rise to some new aspects of the subject.

The more or less generalized argumentations among the frequently rather trying technical discussions must be taken for what they are: attempts to illustrate a situation. The sections concerning the development of life in the first phases may be hazardously subjective; but this is inescapable, since the student of these archaic epochs does not have many facts and data available for a strictly objective description of a milieu but must fill in the gaps with somewhat fanciful assumptions. It is possible that sometime in the future—after a century or perhaps two—a clear and objective survey of the development of life on our earth can be published. At present we cannot expect more than something that may possibly be characterized as an attempt toward a hypothesis.

June 18, 1960

CONTENTS

I. PERSPECTIVE

Life seen as a whole and in detail. An attempt to understand causal relationships. 1

II. AT SOME DISTANCE

Data concerning the elements that in various combinations comprise the different units of life. Metabolism of these elements on a large scale. Development of a discussion concerning the two poles of activity within life: photosynthesis versus respiration. 11

III. CLOSEUP

A brief survey of the chemical metabolism of living cells. Their decomposition and utilization of nutriments. Their energy metabolism and possibilities of resynthesis of cellular material. Their genetic stabilizing factors. 24

IV. OVERVIEW

Discussion concerning the possibilities of understanding the origin of life on our earth. 53

V. FAR AWAY

Development of life in the Cambrian period compared with corresponding life in our time.   60

VI. BEYOND THE HORIZON

Possibilities of determining the age of the earth with the help of isotopes and establishing the occurrence of certain organisms in the pre-Cambrian period.   71

VII. OUT OF DUST AND FIRE

Data concerning formation of the planetary system. The nature of the earliest carbon compounds on the earth.   86

VIII. PATIENCE

Conversion of the earliest carbon material. Origin of the sea. Concentration processes.   99

IX. COMPLICATIONS

Formation of early polymers; their conversion during different geologic processes.   107

X. ACTIVITY

Development of centers for polymeric activity. Occurrence of units with certain characteristics of life. Small crater lakes and their chemical possibilities.   114

XI. VITALITY

Further development of the chemical possibilities in isolated crater lakes. Formation of smaller units within the whole. The role of hydrogen sulphide and iron salts in early metabolism.   123

XII. SOMETHING NEW UNDER THE SUN

Disappearance of the organic substrate. Possibilities of using carbon dioxide as a substrate with the addition of energy. This addition hydrogen sulphide in the beginning, later solar energy. The first photosynthesis.
130

XIII. RECOGNITION

Discussion concerning the formation of the first hereditary factors in early organisms. Their development in the pre-Cambrian period, especially the formation of shell structures. Reflections concerning the development of life-as-a-whole. 138

XIV. COMMENTS

Literature, facts, data, and a short concluding discussion. 147

INDEX AND GLOSSARY

Index and Glossary 157

CHAPTER I
PERSPECTIVE

What is the ideal distance for observing a cloud? From far away, or close up?

A mountain climber on a lofty peak sees a compact, rounded gray and white shape drifting slowly toward him. It seems clearly outlined against the blue sky, an uncomplicated mass, obviously related to sea and warmth, to water vapor rising from the earth.

As the cloud comes closer, its contours begin to disintegrate into a diffuse transition between blue haze and gray turbulence. Twilight sets in, fog sweeps over the slope, and the lonely wanderer—if he happens to be in the mood—can ponder the possibilities of reconstructing an objective, over-all picture of the complex something that now surrounds him.

Somewhat the same situation confronts anyone who attempts to understand the life that surrounds us in all its turbulent capriciousness, the complex something in which, moreover, we ourselves participate as units for brief periods of time. To analyze such a situation requires a certain ability to see things from different perspectives. Where our view of life-as-a-whole represents one extreme, the chemical activity that takes place in those units of matter that we define as living is the other. Both the over-all picture *and* the close-up are used here as starting points to establish a concept of life. The further analysis will show how the two aspects are linked together in a complex whole: how the development of life-as-a-whole on our earth influences the activity of its participating units, and how the latter influences the totality—that is, life in all its manifestations and complexity as it developed from a misty past with a seeming inevitability even to its utmost complexity.

Seen from afar—as observed from outer space, for instance—life on our earth appears as cyclic activity within something that can be characterized as a thin covering around the planet—a constant interaction of sea, land, and atmosphere, especially around the temperate regions of the land surfaces. The main impression of the whole is of something green, distributed in a spotty pattern over the planetary surface—sometimes densely, sometimes thinly—interspersed with regions of brownish red and, toward the poles, clear white. A sharp-sighted observer armed with some technical resources can observe that in certain regions, especially in the markedly green ones, a considerable absorption of carbon dioxide is taking place, which seems to be connected with radiant solar energy. Certain other centers, most often in the non-verdant regions of the earth, seem to give off considerable amounts of carbon dioxide. Since the carbon dioxide content of the atmosphere seems to be reasonably constant and, similarly, the active covering around the planet seems to have the same mean density year after year—with certain small variations—he concludes that an interaction is taking place between regions of different chemical trends.

From new observations, new data are obtained concerning this mysterious flux of carbon dioxide. It looks as if the green regions grow as a consequence of carbon dioxide absorption and as if this growth is effectively inhibited by the activity in the non-verdant regions—a sort of constant decimation, leading to a balance between the centers that release and those that absorb carbon dioxide. At the same time it can be ascertained that not only carbon dioxide but also water is involved in this activity, that the solarized green regions take up both carbon dioxide and water in their growth, and that their counterparts return both components to the atmosphere and sea. The high content of free oxygen in the terrestrial atmosphere—which is unique in the planetary system and probably would puzzle observers unfamiliar with the local conditions—may gradually be seen as a consequence of the incorporation of carbon dioxide and water in the green regions. The oxygen seems to be a by-product, which, however, is effectively taken up by certain centers in the non-verdant regions as a factor in the breakdown of the greenness into carbon dioxide and water.

A distant observer with a certain knowledge of chemistry should by this time be able to sum up a few things. The formation of oxygen from water, and the incorporation of water and carbon dioxide into some indefinable complex are both processes that require a certain amount of energy. It is obvious that the incoming sunlight constitutes the source of energy for these processes; the planet's own energy resources seem too limited to promote the reaction. Thus, in the green regions there must be something that has the ability to convert solar energy into chemical energy, something that mediates the utilization of incoming radiation in the reaction whereby the simple and uncomplicated are converted into the complex—a reaction that has a natural tendency at every favorable opportunity to run in the opposite direction. The high oxygen content of the atmosphere makes it quite clear that the decomposition of the complex—which is evidently composed mainly of carbon, hydrogen, oxygen, and nitrogen—into carbon dioxide, water, and other inorganic units is a process that involves the assimilation of oxygen and the liberation of energy, largely in the form of heat radiation.

The total impression is that around the surface of the earth there exists something resembling a complex catalytic material which is decomposed and resynthesized during the conversion of incoming solar energy into outgoing radiant heat; a complicated chemical system, the purpose of which—seen from a distance—appears to be merely constant consumption of high energy and production of low energy: a transformation of solar yellow to infrared.

It is obvious that anyone who has gradually become familiar with the outward aspects of this extensive chemical activity around the planetary surface will raise some questions about the detailed formation of the whole. First, what is this remarkable catalytic material that mediates all this chemical activity, apparently both in the sea and on the land? How was it formed, and how does the interaction of its components function within the framework of the activity as a whole? What is the composition of these units whose total chemical activity must represent the function and development of the whole? Is there any meaning or import in this enormous metabolism of carbon compounds in constant synthesis and decomposition, this complex conversion process, whose driving force seems to be incoming solar energy and whose purpose seems to be the conversion of this energy into heat and chemical activity?

In asking these questions we relinquish the over-all, simplified view of life around our earth and embark upon a study of the activity close up. Our first impression is of a rather bewildering complex of events, a hive of incomprehensible activity. The simple contours of life that we observe at a distance have changed into a stippled pattern of innumerable interacting units of highly complicated structure that is extremely difficult to understand. Here we come into contact with life-in-detail, life at the level of organisms and cells in hectic chemical activity, within which we ourselves are included as units in a complicated process of material in motion and constant transformation. In all this turbulence we begin to discern the individual forms of the surrounding units, the individual structures of the representatives of the plant and animal worlds, the opposite poles in the activity on a large scale. Our impression of each unit in its typical, specialized form and pattern of activity is that they all have a common, characteristic feature of autoactivity, something that we find difficult to define but designate for the sake of simplicity as *living*.

What we comprehend as autoactivity in cells and organisms and intuitively understand as something living has a marked ability to evade our attempts to analyze its nature more closely. The more we try to penetrate the question of what it is in a living cell or organism that gives it this special character of living, the more we realize how difficult it is to find the correct way of expressing what we obtain in our analysis. The reason for this lies naturally, to a certain extent, in the limited functions of our brain per se, but perhaps still more in the fact that our own role in life is that of both observer and participant. We are thus forced to judge the formation of the living in all its variability from two viewpoints, partly as something that is an extreme expression of the innate tendency of material to complexity carried to perfection and partly as something that in all its outer expression has much in common with our own reactions in a world filled with complications.

We grasp the latter aspect in the situations where we, with a certain personal involvement, observe an organism in full activity, a flowering plant or an animal in its natural milieu. Life on the organism level appears to be so self-evident! We see something that takes up nutriments, maintains the integrity of its structure by conversion of the nutrient components to organism material, gives off waste products, moves slowly or rapidly, exists for a short period, and later, through external or internal circumstances, undergoes destruction as an individual unit. Something that during its short period of existence appears to have a purposeful drive, self-initiating activity; something that seems to have its being *in order to* rather than *because of*. When this autoactivity on the individual level reaches its natural end, we have a situation we call death: the contrast between the inactive stuff and the earlier active; between material in dissolution and the earlier *élan vital* of the whole; the withered plant in contrast to the flowering; the dead fish versus the living—something self-evident and comprehensible, a reflection of our own situation in a pitiably short existence of activity, goal-seeking, and resignation.

Consciously or unconsciously, we ascribe a motive force to all this activity which is localized in the cells and organisms themselves; we invest with an active initiative manifestations of life of all kinds which have their origin in the individual unit—so long as it lasts. At the same time, however, we have a definite feeling that all this activity has a connection with another aspect of life, which in passing we defined as material's innate tendency toward complexity. From this latter point of view, life, with its manifestations on the organism level, has a background that in some way is impersonally chemical, like the activity within life-as-a-whole. The driving force for the activity lies outside the whole: incoming solar energy, inciting activity in units adapted to accept it. The rest consists of consequences, whose outer forms of expression in interaction with the other units of life give us a highly illusory impression of purposeful activity.

What we comprehend here as cells and organisms are expressions of a special organization of material which has been carried so far that each unit has acquired individuality; this becomes evident in the interaction of the units. Each unit is individually sensitive, but the reactions to a certain stimulus are more automatic than directed by an individual motive. What we see of the activity of a living organism is a manifestation of complicated reactions of these cells, constant conversion of nutriments to cellular material and waste products, a continual stream of transformed material through the organizations that are cells and that are maintained as cells by this stream of material in constant transformation.

In all this we can learn by chemical methods how carbohydrate, fat, and protein are synthesized and decomposed in a series of complicated chemical reactions, how energy is stored as reactive phosphates, how the whole cellular organization is influenced by genetic factors, how the organization can maintain its structure intact in a fluctuating environment and, in addition, at times propagate its own species through cell division: the formation of two new organizations with the same chemical plan. The more we see of a cell in its automatized perfection as a transitional stage for a stream of material, the more we wonder how this impersonal thing, this center of chemical activity, represents life according to our own personal definition of the concept of life.

It is undoubtedly life on the cell level—so far we can agree. But, at the same time, the automatism gives it more an aspect of the *élan chimique* than of the *élan vital* which we thought we discerned in the cellular organization observed as an organism. The concept of life which seemed so self-evident—although indefinable—when we saw an organism in full activity appears during the course of the analysis to have slipped through our fingers when we began to examine a cell in detail. We found a wonderfully well-organized chemical activity, which made a cell what it is, an activity center for the conversion of the surrounding abundance of chemicals into the cell itself, plus something else. We found that the cell in all its intricate organization is only a transitional stage for a flux of material, that its existence as a cell possesses little initiative, that its brief existence as an individual unit is entirely dependent on the supply of nutriments. At the same time we can ascertain that, owing to its chemical nature, the cell's maintenance of organization and its eventual division into two cells give it a chance, in an environment containing nutriments, to transform these into cellular material and to perform all the functions that are characteristic of the living cell in contrast to the non-living.

This is a dilemma. In continuing our analysis and attempt to obtain an answer to the question *why* a cell is what it is, we find that the chemical linkage pattern underlying all that we can observe of cellular structure and function has a developmental history. A cell is a chemical organization which can be derived from another cell, which in turn has developed from another, and so forth. All this development, this continuity, this chemical tradition in regard to the transformation of material in the surroundings into some definite micro-organization dates from an early period in the evolutionary history of the earth during which the rules for what we observe as cellular chemical activity were formulated; a period when something was evolved which—transformed during millions of years—became what we today comprehend as cells in every activity, in individual operation, and in complex co-operation as organisms, lower and higher.

If we try to understand the motive force, seen locally, of cellular growth and division, of both now and millions of years past, we find that the answer in most cases is a stern statement of facts: other cells, other organisms. The driving force of any cell in its operation as a cell depends upon a certain concentration of nutriments around the cell surface. In the particular case of green plants, it is the accumulation of carbon dioxide and water around specially organized units in green leaves; in most other cases it is the debris of cells which have met with some environmental calamity, possibly among more complicated waste products of cellular activity. It is, on the whole, an automatized situation, where the joint action of the chemical organization within the cell and its surrounding environment of chemical components *must* lead to a characteristic result: first, the maintenance of the organization for a certain period and thereafter either death through decomposition or re-formation into two new organizations so long as the chemical conditions for the interaction between the cell and the environment permit this development. There is an old formulation for chemical reactions in general that says quite unemotionally: If there are in a solution two components which under given conditions can form an insoluble compound, even though with difficulty, *then this is formed*. Transferred to the infinitely more complicated interaction between the cell and the environment, the same formulation can be expressed: If a cell can form its own cellular material from a number of chemical components in its environment, *then cellular material is formed*. The same strict consequence is also valid for the maintenance of the cell as a cell in a certain environment, its death as an individual through decomposition, and its reorganization into two new cells through division: If conditions permit, it happens!

The variations on this theme are innumerable. Whatever furnishes every cell or cellular organization with what it needs to absorb of nutriments, to grow, and to propagate is evidently nothing more than a local environment rich in other organisms or chemical substances. An abrupt environmental change, and the cell ceases to be a cell, the organism to be an organism. Within the framework of an enormous cycle, plant and animal life, water, carbon dioxide, and other vital units are incorporated into a system of stimulation of all by all, through disintegration and reorganization. The purpose of every unit—organism, cell, or chemical compound—is to be itself for a short period and then to become something else.

> Theirs not to reason why,
> Theirs but to do and die.

From a broad viewpoint, all this is a simple consequence of solar energy radiating on our earth and slowly reacting water and carbon dioxide being constantly activated to synthesize the complex molecular combinations of plant life. This material, after passing for a certain time through the chemical labyrinths of the animal world, is eventually found again at the lowest energy level as regenerated carbon dioxide and water, ready for new activation, new infusion of solar energy mediated by the green plants. It is the cycle of energy consumption that we observed from afar, the apparent purpose of which is just to be this cycle, the individual contributions within the framework of the whole representing something fortuitous, yet forming an inevitable series of single events within a system of causal relationship.

This is life, in detail and in general—a somewhat impersonal system of carbon compounds in constant transformation, within which our own development constitutes a specific event which occurred during the most recent moments of a long history of events. But here, for some reason, the tendency of matter toward complication has been carried so far that some organisms have attained that degree of complexity which is reflected by advanced consciousness. This "something" creates a desire in us to point out our unique position at the end of a course of evolution; it underlies our ability to comprehend time and to be able to see things from different perspectives, giving us the opportunity to understand something of the logic of life, both in the earliest stages and in our time. As an automatic consequence of a development of the potentialities of carbon compounds, this is worth noting.

Every attempt to analyze life and its history tends, we see, to avoid any definite answers to our questions; or, more correctly, the answers are not exactly those we expected. The answer to the question, "What is the true nature of life?" is that it is what it is; and it appears to be, in simple outlines, a cyclic procession of matter driven by sunlight. Our attempts to obtain from cells of different types any deeper secrets, locked in the machinery of cell plasma, organelles, and chemical substances of all kinds, yield only obscure answers, which seem to suggest that the cell as a living unit is a consequence of the life and ways of living of other cells, and thereby a consequence of the whole system, which we can hardly comprehend as something living per se. Moreover, we ourselves have developed as organisms with the capacity to ask questions, a characteristic that is sufficiently complex to withstand most analyses.

What we can do in this situation is patiently to utilize our newly acquired ability to grasp the import of different events from different perspectives, making use of our unique position in life as observers *and* participants. It is possible that if we gradually collect data on the formation of life-as-a-whole as well as in its detail and endeavor to trace its development from the dawn of time to the present, we shall become more familiar with what appears to be an enigmatic manifestation or a manifest enigma. We may gain a little understanding of some of the happenings in our surroundings which seem incomprehensible and learn to appreciate some new facets in the structure of the whole. Naturally it is not necessary, but it may entail some effort, some pleasant but not profound mental exercise. It is like observing a cloud against a blue sky which gives the appearance of clarity and changeableness and stimulates reflections on certain aspects of life.

CHAPTER II
AT SOME DISTANCE

A suitable starting point for analyzing life-as-a-whole from our observation post in the present is to concentrate on the general features of the reshuffling process that give present-day life its characteristic structure. We can introduce facts in the form of figures without involving ourselves to any great degree in theoretic speculations. To begin with, we can state unemotionally that present-day life comprises three reservoirs of material: *first*, inorganic matter, of which carbon dioxide, water, and some nitrogen, sulphur, and phosphorus compounds are the main components; *second*, all green plants with the capability of forming from the material in group I the material of their own group II; *third*, myriads of animal units, all hungrily consuming group II and each other and thereby supplying carbon dioxide and water to group I. If we wish to estimate the quantitative relationships in the different groups, figures are available in the form of the amount of carbon in the different reservoirs. Employing this distribution of carbon within life-as-a-whole is a purely practical measure, in that we can obtain at the same time an idea of the annual turnover within the entire system.

Let us glance at the table on page 12.

The calculation of tons of carbon in the animal and plant worlds is uncertain, but it lies within the magnitudes given.

As a complement to this table, which with its dry figures resembles the balance sheet of a large corporation, we present a more abstract diagram in Figure 1, in which the quantitative relationships of the different groups are symbolized by their respective areas. In the scale employed here 1 mm² corresponds to 1 billion tons of carbon. See page 13.

From the above table and the diagram we can ascertain the following: the carbon material in group I is mainly localized in the sea; the carbon content of the atmosphere is relatively slight, the carbon dioxide concentration of 0.03 per cent yielding hardly 600 billion tons of carbon, which is a relatively small fraction compared with the 50,000 billion tons of carbon dioxide and carbonate carbon localized in the sea. However, such a sensitive equilibrium exists between the atmosphere and the sea that the annual withdrawal of ∼15 billion tons of carbon in the form of carbon dioxide from the atmosphere's 600 billion tons is rapidly replenished from the enormous reserves of the sea. Furthermore, we can ascertain that the annual conversion of ∼30 billion tons is drawn equally from the land and sea regions, which at first glance seems to contradict

		Billion Tons of Carbon	Total
I	Carbon dioxide in the atmosphere	∼ 600	
	Carbon dioxide and carbonates in the sea	∼ 50,000	∼ 50,000
II	Plants		
	Land	30–60 }	∼ 50
	Marine	3– 6 }	
III	Animals		
	Land	1– 3 }	∼ 5
	Marine	1– 3 }	
IV	Carbon temporarily withdrawn from the present turnover:		
	Carbonates, land	1,000,000 }	∼1,000,000
	Fossil non-carbonates	10,000 }	
	Annual conversion		
	Photochemically bound carbon/year (newly formed plant material):		
	Marine	∼ 12–15 }	∼ 30
	Land	15 }	
	Energy addition in the form of sunlight utilized for synthesis of plant material	per year: ∼10^{19} kg/cal.	

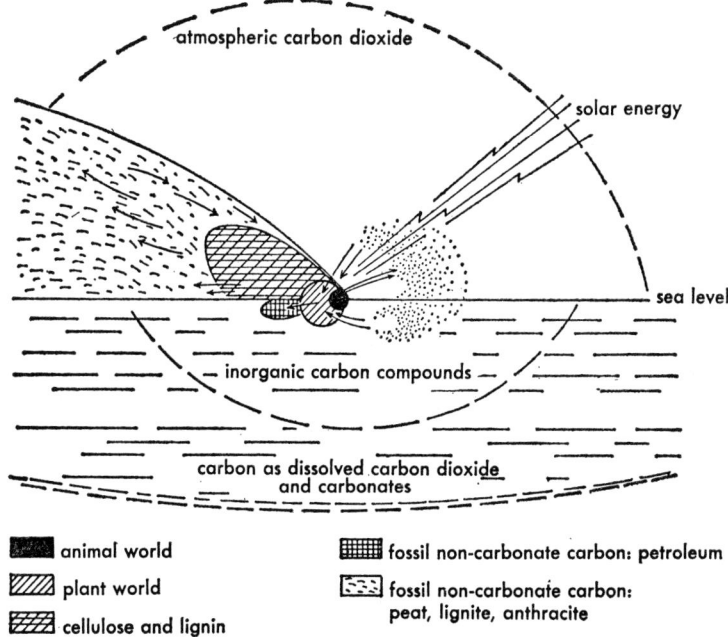

Fig. 1.—Schematic diagram of life-as-a-whole. (Arcs and semicircles should be conceived as fully described.) Each mm^2 represents approximately 1 billion tons of carbon in different combinations: as plant material, animal material, fossil non-carbonate carbon, and carbon dioxide in the sea and atmosphere. The distribution within sea, land, and atmosphere corresponds to that given in the table on page 12. However, this figure shows a dynamic flux: constant growth of the plant world from carbon dioxide and water, a process driven by sunlight, and the animals' effective consumption of the plant life and one another, with consequent re-formation of carbon dioxide and water. At the same time we can see that a small fraction of the plant world is excluded from the general process; namely, the most indigestible, cellulose and lignin, mainly wood. Some of this material gradually becomes the deposits of fossil non-carbonate carbon which we call anthracite and petroleum. It is at present undergoing combustion at such a rate that deposition can by no means keep up with the intensive decimation of industrial activity. Man—in billions of tons of carbon—is equal to a tiny dot within the black region, about 0.5 per cent of the total animal life, an effective catalyst within the whole.

the fact that land plants predominate over marine plants by approximately 10:1. Actually, this is due to the fact that the main part of the carbon material that represents land plants is bound to wood, to the cellulose and lignin of trees and shrubs; the units responsible for the carbon metabolism—the green parts containing chlorophyll, mostly leaves—represent, so far as land plants are concerned, only a small per cent of the total material, whereas the chlorophyllose parts of the green marine plants comprise a considerably greater part of their structure. The ratio of 1:1, referring to metabolized carbon per year from sea versus land, reflects roughly the quantitative distribution between sea and land of units containing chlorophyll within the plant world—between the green algal flora of the sea and the green leaves and stalks of the land regions.

Another fact which is rather surprising is the unusually rapid turnover within the system. Thirty billion tons of atmospheric carbon are utilized by 50 billion tons of carbon in the plant world, to be consumed in turn within the year by 5 billion tons of carbon representing an ever-hungry animal world. The capacity of animals to consume plant life and each other is five times their own weight annually, calculated as carbon—an appetite worthy of respect. A small fraction of animal life, representing the not insignificant amount of 0.02 billion tons of carbon—we humans—in addition to consuming annually about 0.1 billion tons of carbon as food, can, moreover, be observed to enrich the carbon dioxide and carbonate stores of the total system by an occupation that is quite novel for living organisms. It can effectively consume fossil carbon material in the form of anthracite and petroleum by means of industrial activity and convert it into carbonate at a yearly rate of 2 billion tons—twenty times more than the nutriments it ingests. This activity of *Homo sapiens* is something new for life-as-a-whole. Broadly speaking, man's special activity is using the resources of carbon material that have been accumulated through fossilization of earlier organisms during the last five hundred billion years. It is, so to speak, capital being withdrawn from the firm and rapidly circulated in our dazed delight at having found readily disposable funds, for the annual withdrawals correspond to ten thousand years' deposits—a magnificent and dazzling extravagance.

In order to make a more precise analysis of present-day life-as-a-whole than this first superficial presentation, some attention should be devoted to the process in green plants that converts carbon dioxide into organic material. It is quite evident that this is the motive force of the whole system: the utilization of solar energy to raise energy-poor carbon dioxide to the level of organic material. The rest of the great cycle represents largely the consequences of this initial reaction: a constant impoverishment of the formed plant material through the destructive influence of animals and people through what we call breathing, respiration, the antithesis of photosynthesis—carbon-dioxide fixation. We can formulate the two processes in a simplified manner:

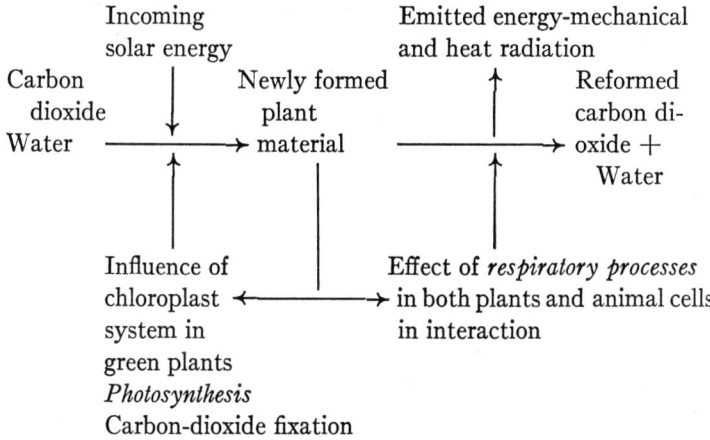

This diagram is a brief summary of the main features of the complicated conversion process that constitutes life-as-a-whole; it is a summation of all new synthesis and decomposition in all cellular material on our earth. In order to get an idea of the enormously complicated processes involved in the formation of cellular material from inorganic substances, we can try to analyze the main reactions that take place in green plant cells when solar energy acts on the combined carbon dioxide and water.

The role of water as a partner to carbon dioxide, shown greatly simplified in Figure 2, has been recognized only in recent years. Seen as a whole, the process involves the splitting of water, H_2O, effected by sunlight, into its components, hydrogen and oxygen, whereupon the hydrogen is bound to certain hydrogen acceptors (symbolized as X in Figure 2) and oxygen is released. The carbon dioxide, on the other hand, is bound to certain carbon-dioxide acceptors (symbolized as Y), forming carbon-dioxide-Y, which is later *stabilized* by hydrogen X, forming something which in several complicated reactions is converted into a still more complicated form—the material of the cell where all this takes place. In this case it is a plant

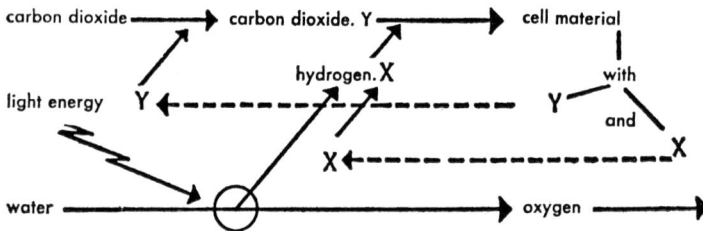

FIG. 2.—Symbolic diagram showing certain main chemical tendencies within life-as-a-whole. Carbon dioxide loosely combines with something we can call a carbon dioxide acceptor, Y, a substance present in all green plant cells all over the world. At the same time water in these cells is split into oxygen and hydrogen, whereupon the hydrogen is bound to certain hydrogen acceptors, here called X. The combination of loosely bound carbon dioxide with bound hydrogen, hydrogen X, leads to the formation of the complex we call cellular material, in this case to growth of green plant cells and to growth in general. Oxygen escapes to the atmosphere and the sea.

cell with specialized organization of chlorophyll and acceptor substances X and Y and innumerable similar substances, each with its own function.

This is what happens every moment in all green plant cells exposed to an obliging sun: the splitting of water into its components, fixation of hydrogen and carbon dioxide, stabilization of bound carbon dioxide with bound hydrogen, and further conversion in the plant cells of the new combination into material that constitutes the cells themselves.

Seen from this point of view, the enormous cycle of the migration of carbon—from carbon dioxide to plant life to animal life and back again to carbon dioxide—is at the same time a migration of the components of water. Actually it is the combination of carbon dioxide and water which, constantly activated by energy in the form of sunlight absorbed by green plant cells, yields organic material. The oxygen is a by-product, enriching the atmosphere and sea. At the same time we discern something else: the reverse mechanism, the destructive phase in the second act of the process, symbolized in Figure 3. In principle, this reaction is the mirror image of the syn-

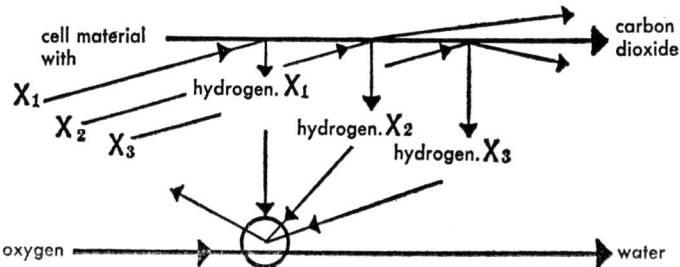

Fig. 3.—The reversion of the complex to the simple through the action of animals on plant material and themselves. Hydrogen acceptors are charged with hydrogen from the nutriments and combined with oxygen to form water. Carbon dioxide is split off from the material in dissolution and disorganization. The cycle is completed; water and carbon dioxide again undergo complex formation, according to the scheme shown in Figure 2, and later are reduced again to the lowest energy level, all promoted by the solar energy that splits water in green plants into oxygen and a hydrogen-charged something.

Figures 2 and 3 are summarized in Figure 4.

thesis mechanism. All cells that are capable of consuming organic matter have a tendency to split off from it bound hydrogen, which is symbolized in the diagram as hydrogen X_1, X_2, X_3. . . .

Under the influence of this successive splitting-off of hydrogen, the carbon compounds of the material begin to disintegrate, step by step, liberating carbon dioxide as an inevitable consequence. The acceptor-bound hydrogen is combined in a complicated sequence with oxygen from the sea and atmosphere, resulting in water. The respiratory processes throughout the world thus restore carbon dioxide and water for the benefit of plant life, within the great cyclic system of consumption. And the whole process continues.

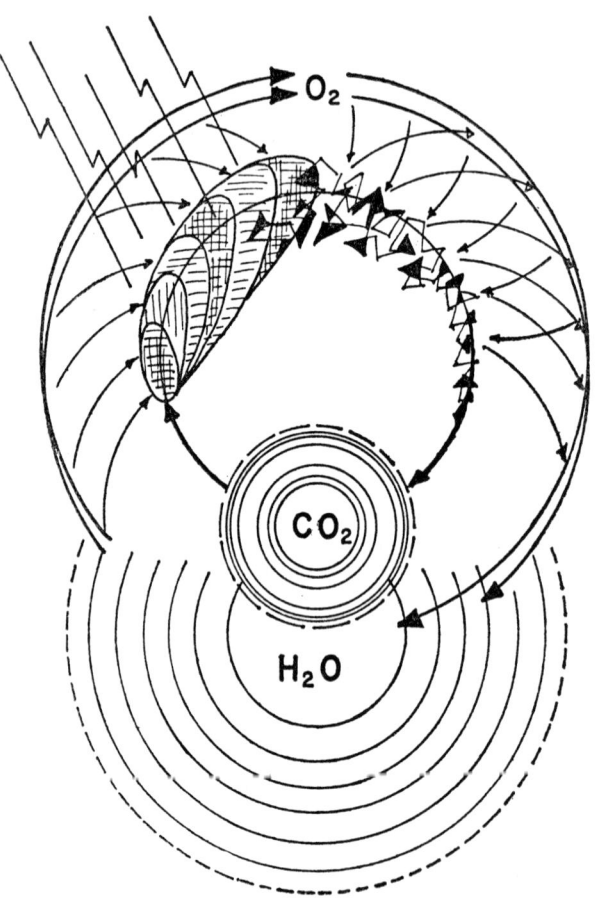

FIG. 4.—Summary of the diagrams shown in Figures 2 and 3. Growing green plant life is consumed by the animal world in a series of respiratory processes. H_2O = water; CO_2 = carbon dioxide; O_2 = oxygen. All this moving around is the consequence of the impact of solar energy.

This interplay between complication and simplification, between synthesis and decomposition, involves—in addition to carbon, hydrogen, and oxygen in different combinations—a number of other components in what, for the sake of brevity, we call here "cellular material." In the first place, we have nitrogen, sulphur, and phosphorus, which are ingredients of the complex structures that constitute protein, the components of the enzymatic machinery of the cell. Furthermore, there are various elements in minute proportions that exercise extremely important, specialized functions in cells of different types. However, what one might term the fundamental material factor within the entire system of all these elements, in continuous passage through the diverse units of life, is something that resides in the tendency of the carbon atoms and, consequently, the carbon compounds to create complications. Carbon's unique quality is that its combinations with hydrogen and oxygen have a tendency in cells of different types to incorporate a number of other elements, forming innumerable new combinations of the most varied structures. When we consider the annual turnover of different elements within the system of life-as-a-whole on our earth, as shown in the following table, we see how the potential combinations of carbon compounds with other elements can create the profuse variation represented by living organisms within the framework of the whole:

ANNUAL CONVERSION BETWEEN INORGANIC
AND ORGANIC MATERIAL

	Billions of Tons
Carbon	$\sim$ 30
Hydrogen*	8–12
Oxygen	$\sim$ 100
Nitrogen	2–6
Sulphur	0.1–1
Phosphorus	0.1–1
Silicon	$\sim$0.1
Iron	$\sim$0.01
Copper	$\sim$0.001

* Hydrogen undergoing hydrogenation and dehydrogenation processes.

Our picture of life-as-a-whole is beginning to be complicated. From a simple outline of the flux of organic carbon compounds in continuous synthesis and decomposition, seen against a changing background, we begin to discern a network of cyclic processes intertwined in a complex pattern: the metabolism of carbon, hydrogen, oxygen, nitrogen, sulphur, phosphorus, and all the others. Our impression of the whole is perhaps still influenced by the simple contours, without details—an impression that what takes place is what appears to take place. At the same time we begin to experience a definite feeling that something in the innermost structure of matter is a prerequisite to the whole's existing as it does; that a co-operation between matter and energy in dynamic interaction from the beginning of time gradually complicated matter, complicated its conversion patterns, complicated its co-operation with different forms of energy, in turn leading to new complications, and so forth.

This should not be taken as a profound statement, but rather as a laborious attempt to summarize the essentials of life without using chemical symbols. To alleviate the situation somewhat, we can for a moment ponder the nature of the special factor in the material of life-as-a-whole that gives the components their properties of variability, convertibility, versatility, and sensibility within the framework of the whole.

Here the question leads back to something we noted in passing regarding the peculiarity of carbon atoms. Carbon has a unique position among the elements of the universe—something that simply is so and must be accepted as a legacy from the formation of the elements in the first seconds of the universe. The uniqueness of the carbon atoms lies in their pronounced ability to form carbon-to-carbon configurations—chain structures, ring structures—often with a variety of component atoms, within which hydrogen, oxygen, nitrogen, and other elements have a unique chance of being incorporated to make structural patterns of infinite variation.

The basis of this tendency toward complex molecular formation is a marked stability in energy pattern—the bonds that link carbon atom to carbon atom, and carbon atom to hydrogen atom, to oxygen atom, and to several other elements. No other element can arrange its atoms in such stable configurations as carbon; no other element can incorporate other atoms into its molecular structure with such great variation in pattern. Only in molecular structures of carbon in combination with itself and with oxygen, hydrogen, nitrogen, sulphur, phosphorus, and other elements do we find the multitude of conversion possibilities, the chemical versatility, which we observe in the myriad of components that constitute life of today and yesterday.

It is just this reactivity of carbon compounds in general that is displayed in the great interplay of synthesis and decomposition: the ability to be grouped together in different forms, to become complicated, to be converted, and, under certain conditions, to be split up and divided, to degenerate to the simple from the complex, and again to be reactivated and reinvolved, to become something else.

However, for this tendency to be activated and to develop freely in all variations of the theme, the process must be initiated by and have constant access to energy in certain regions of the arena. A further requirement is an environment in which temperature and pressure remain within certain definite limits and give decomposition and death a clear chance to counterbalance the energy-initiated synthesis. If it is too hot, the material will never exceed a certain degree of complexity; it remains at the level of molecules with few reactive atoms, active but without a chance of developing into high molecular vitality. If it is too cold, the process will slow down, in some cases with the result that synthesis outruns decomposition and death. If unchecked synthesis should combine the uncomplicated with the complicated without the intervention of destruction, we should, after a certain time, arrive at a static situation, a final result representing the perfected tendency of carbon compounds to complications—if that is at all conceivable. Some sort of inert material or some slowly reactive guano, whatever we desire to call it, would cover the planetary surface with impressive monotony. To sink into such a situation, the originally existing carbon compounds of a low degree of complexity would have to be exposed to such a one-sided environmental influence, including energy additions of different kinds, that no decomposing fatalities would affect the more complex material. The result would be a final product somewhat inert, congealed, with limited chances of being transformed, via destruction, into activity again.

Very possibly something like this situation exists on some of our neighboring planets. Venus' dry, carbon-dioxide-saturated environment and Jupiter's cold surface of methane and ammonia are probably sterile situations where something that once occurred no longer occurs; a complex crystallization of possibilities, and then nothing more. The final result is self-sufficient.

A tendency in the same direction can be observed on our otherwise rather dynamic earth. In the process of life during the last 500 million years an imposing amount of carbon material has been withdrawn, so to speak, from the process-as-a-whole—everything that in the table on page 12 is designated as fossil carbon, temporarily withdrawn from present conversion. It is partly simple and uncomplicated material—carbonates surrounded by silicates, representing skeletons and sheaths of prehistoric organisms, now incrusted in bedrock as calcareous and siliceous fossils—and partly highly complicated material—anthracite and petroleum, earlier organism material buried beyond the reach of fellow organisms' appetite. Man's entrance into the arena has been followed by a brutal awakening of some material from its chemical sleep. Limestone is turned into lime and carbon dioxide; anthracite and petroleum likewise gradually become carbon dioxide, with the chance of repeating the cycle as plant and animal life during constant complication and simplification, activated from a more or less static condition into a pronouncedly dynamic one.

Destruction through constant degradation of the complicated appears thus to be absolutely essential for the synthetic processes of life to function and thereby keep the system running. From this point of view, synthesis and death are partners in the greater whole that is life. We humans possibly comprehend this as expedient reality when it concerns nature around us, but as grim determinism when it concerns us.

This is what we grasp when we look at life-as-a-whole, life from a distant perspective. From this wide-angle view, the function of life appears to be conceivable in data, figures, facts, but at the same time there is something puzzling and inaccessible for the person who tries to raise questions about the individual organism. The questions

"What is life?" and "How has it originated?" often have such emotional overtones that an answer referring to the whole and its development in all its chemical complexity is unsatisfying.

We shall now examine something more living than life-as-a-whole and see what we can learn from study of an organism under the microscope.

We know in advance that again we will gradually arrive in a world of chemical conversion processes—this time on the level that is represented by living cells. Our penetration of the chemical reactions within the cellular membrane will clarify the laws for these reactions but will tell us hardly anything about what produces the vital individuality we unconsciously attribute to bacteria in action, the lilies in the fields, the ant in the ant-hill, the lark in the sky, and the writer at his desk. There is a destructive element in all analysis —we know this so well that comments are unnecessary—but we also know that the results of an analysis, recombined to a new synthesis, to a total picture, can enrich our view of certain things. From a quick study of life-in-detail as cells and cell organizations, we can again view life from a distant perspective, perhaps with something gained, perhaps with a starting point for obtaining a glimpse of its origin.

CHAPTER III
CLOSEUP

Under the microscope we observe tiny swimming objects: oval units, sometimes in pairs, sometimes alone, occasionally with unbroken contours of indefinite color, now and then with an opalescent globule as an adornment to the spare simplicity. It is a little colony of cells, some dividing, others still single but with a good chance of gradually undergoing the chemical convulsion that makes one living cell two. The opalescent globule is a bud, a beginning that will grow out: something new, but invariably the same type as the mother cell. Inside the environment are nutriments; outside the environment is something that for some reason provides the cells with their special nourishment. This is a routine picture in innumerable laboratories throughout the world.

The eye above the microscope discerns growth of the cellular material on this special nutrient medium, this special substrate which stimulates the cells to take up food and transform it into the constituents of more cellular material until gradually two individuals are formed where previously there was one.

The theme is repeated with innumerable variations in the interplay between different types of living cells and substrates of different composition. What we have been observing happens to be a colony of yeast cells of the type *Torula*, distant relatives of the advanced types of the order of *Fungi*, which in a meritorious manner contribute to our breadmaking and, in certain cases, to the brewing of good strong beer, in both cases acting by virtue of their by-products —what is formed by the cells in addition to the cells themselves.

This is our point of departure for the study of a cell in detail: this conversion of surrounding material into cells, plus something else; a small fraction of the ingested nourishment being complicated to such a degree that it becomes cellular material while the rest undergoes simplification beyond recognition and leaves the cell as carbon dioxide and other products. This is a performance with a moral: many are called but few are chosen. The few exist for a while in a highly organized condition at the expense of the others' destruction and then are ultimately simplified and combined with the other units in a joint exit from the stage.

What builds up a cell, and thereby living organisms, into different formations is, as is well known, a complex variation on the theme of the conversion of carbohydrate, fat, and protein. Regardless of whether the supplied nutriments comprise largely similar material— the result of some previous calamity to another organism—or simpler components in special combination, the absorbed material will undergo a *process of normalization* in the interior of the cell before attaining such a degree of chemical aggressiveness that it can be the starting material for the synthesis of cellular substance as such. We know also that this mechanism for simplification of the *multitude* and complication of the *few* functions according to certain rules which are valid for all living matter, without exception.

To attempt to elucidate some of these *principia biochimica* without the help of concise chemical symbols is a difficult task—like trying to give a detailed description of the structural beauty of written Chinese without illustrating it with any graphic symbols. We will by no means completely substitute the shorthand of formulae for cumbersome verbal presentations; possibly now and then we can achieve a compromise.

We can start with the principle according to which carbon dioxide is continually released from cellular nutriments. The main part of these reactions runs step by step, occasionally a pair of hydrogen atoms splitting off, thereby making the molecule increasingly hydrogen-deficient. At a certain stage the molecule becomes so unstable that it has a pronounced tendency to split off carbon dioxide, thereby being converted into something more stable. Let us give the chemical symbols a chance; H represents hydrogen, C carbon, and O oxygen:

$$\ldots CH_2\text{---}CHOH\text{---}COOH \xrightarrow[\;XH_2\;]{\overset{\overset{\displaystyle E_1\;\;X}{|}}{}} \ldots CH_2\text{---}CO\text{---}COOH$$

$$\ldots CH_2\text{---}CO\text{---}COOH \xrightarrow[\;CO_2\;]{E_2} \ldots CH_2\text{---}CHO$$

We have here the substance symbolized as ... CH_2—CHOH—COOH exposed to a hydrogen-transferring enzyme E_1. The result of this reaction is that another substance, which for the sake of simplicity we call X, is made to accept the split-off hydrogen—two hydrogen atoms—thereby forming the combination XH_2. The remaining combination, ... CH_2—CO—COOH, is so unstable that in the presence of another enzyme, E_2, it splits off carbon dioxide, CO_2, thus becoming one unit shorter. The directing influence of the entire process in its two steps lies in what we call enzymes, which are proteins with the ability of directing a reaction merely through their presence; they function as *catalysts*, stimulating a sluggish tendency to reactivity to become a smoothly running process.

Each cell has a more or less loosely organized supply of enzymes, all to a certain degree *specifically set* to effect a certain chemical transformation. The substance to be transformed fits into the architecture of the enzymatic structure in some way which since the day of the German chemist Emil Fischer has gone under the name of the lock-and-key theory. The key in this case is the substrate molecule; the lock is the enzymatic structure. The result of the contact between the key and the lock is that something is opened. On the molecular level this means that a bond between two atoms is opened, or broken, with the result that something is given off—in the above case, two hydrogen atoms (*dehydrogenation*), or a carbon dioxide molecule (*decarboxylation*).

Other enzymes work on other material, often in combination with one another, and with amazing dependability can effectuate decomposition through dehydrogenation of extremely complex material. A longer chain fatty acid, for example, one of the constituents of what we designate as fat in general, undergoes breakdown in a sequence like this:

$$\ldots CH_2CH_2CH_2COOH \xrightarrow[XH_2]{E_1 \quad X} \ldots CH_2CH=CH-COOH$$

$$\ldots CH_2CHC=H-COOH \xrightarrow{E_2 \quad H_2O} \ldots CH_2CHOHCH_2COOH$$

$$\ldots CH_2CHOHCH_2COOH \xrightarrow[X'H_2]{E_3 \quad X'} \ldots CH_2COCH_2COOH$$

$$\ldots CH_2COCH_2COOH \xrightarrow{E_4 \quad H_2O} \ldots CH_2COOH + CH_3COOH$$

in which the original fatty acid . . . $CH_2CH_2CH_2COOH$ is shortened to . . . CH_2COOH. The cleavage product here is not carbon dioxide but acetic acid, CH_3COOH, in a special form. In general, however, the principle is the same: the taking-up of enzymatically split-off hydrogen by certain hydrogen acceptors, here X and X', forming XH_2 and $X'H_2$.

In this system of four co-operating enzymes, we see that water has been taken up in two places and that hydrogen has been given off in two places (i.e., dehydrogenation has taken place). The final result is a splitting-off of acetic acid (in a special form) from the long chain, which is now ready for a new round of simplification. Proteins, in contrast to fatty acids, can be broken down in reactions involving only a taking-up of water; through this *hydrolysis*, long chains of units of the type —NH—CHR—CO—NH—CHR—CO—NH—CHR—CO, etc. are gradually broken down into a number of normalized structures—approximately twenty—of the general form: NH_2—CHR—COOH, called amino acids, the symbol R standing for about twenty different structures of carbon compounds.

So far as the carbohydrates are concerned, we have a decomposition pattern comprising alternately the uptake and the splitting-off of water and a dehydrogenation step, all leading to a normalization product which we designate pyruvic acid, CH_3CO—COOH, from which much can later be obtained:

$$CH_3CO\text{—}COOH \xrightarrow[\downarrow CO_2]{} CH_3CHO \xrightarrow[\downarrow XH_2]{\mid X \quad \mid H_2O} CH_3COOH \quad \text{Acetic acid}$$

$$CH_3CO\text{—}COOH \xrightarrow[\downarrow CO_2]{} CH_3CHO \xrightarrow[\mid XH_2]{\uparrow X} CH_3CH_2OH \quad \text{Ethyl Alcohol}$$

$$CH_3CO\text{—}COOH \xrightarrow[\mid XH_2]{\uparrow X} CH_3CHOH\text{—}COOH \quad \text{Lactic Acid}$$

In this series of formulae in chemical Chinese, where for the sake of simplicity we have omitted the symbols for the different enzymes E_1 through E_4, we see in the top row an example of carbon dioxide cleavage and the reaction product being subjected to dehydrogenation with the co-operation of water—one variant of the theme. In the two lower reaction series a reversal of dehydrogenation occurs, a reaction in which hydrogen-charged XH_2 donates its hydrogen to a compound—a *hydrogenation*. The hydrogen resulting from different dehydrogenation reactions can thus be further utilized for hydrogenation, the latter phase most often being geared into synthetic processes that lead to the complicated substances of cellular material; this is in contrast to the dehydrogenation mechanisms which generally presage a destructive simplification.

To venture a simplified summary, the whole process looks like this:

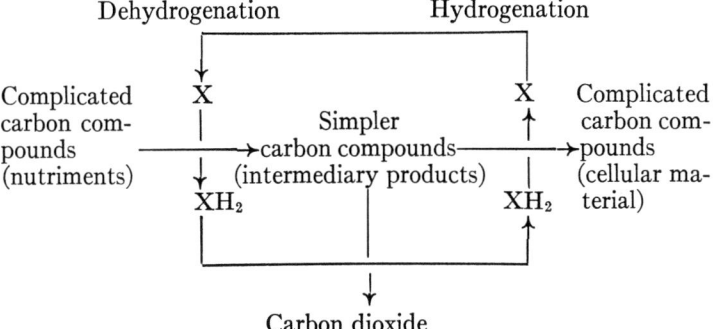

We must add two things to this simple scheme which are quite obvious. First is the fact that all cellular material exists only for a short time as such and is *continually decomposed and resynthesized* within the cell. Second is the fact that in most types of organisms and cells of the group Animalia a certain excess of XH_2 is combined with the oxygen of the air, forming $H_2O + X$. We add these facts to the foregoing diagram:

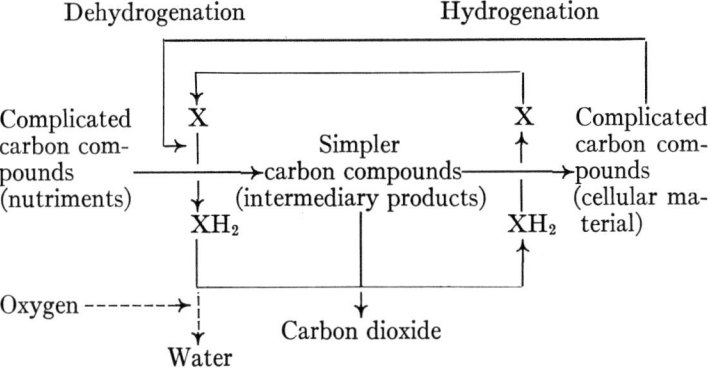

The diagram is becoming more complicated, and we are beginning to meet with difficulties in our efforts to record facts in a simple form. Figures 5 and 6 may help to illustrate something we noted in passing in the first part of the chapter: namely, that of the multitudinous chemical structures of the nutriments, *most of the simplified, the normalized, leave the cell as carbon dioxide and other products.* A small fraction of the remainder are transferred through the action of the hydrogen acceptor, XH_2, to complicated structures within the cell—or, more concisely, become cellular substance. This organization of complicated structures—complexes of fats, carbohydrates, and proteins, and consequently some enzymes—*is likewise subject to the laws of dissolution,* and the components are gradually combined with the simplified, to leave the cell again as carbon dioxide and other simplified products. In an environment containing oxygen, the excess XH_2 leaves the cell in the form of water, reforming the X (see Figures 5 and 6), which is now ready for renewed activity as hydrogen donor.

Now we can consider for a moment the concept of synthesis, which we have formulated rather loosely without due consideration to chemical symbols. We know that, seen as a whole, the formation of cellular material involves *hydrogenation* in contrast to the *dehydrogenation* of the simplification process. One example is the resynthesis of fat, which process is largely a reversal of the fatty acid breakdown outlined on page 27. It is, thus, a complicated coupling of a number of acetic acid units—in this case in a special reactive form; during a gradual splitting-off of water and hydrogenation, long carbon chains are formed from the simple ones. The reaction may be summarized as follows:

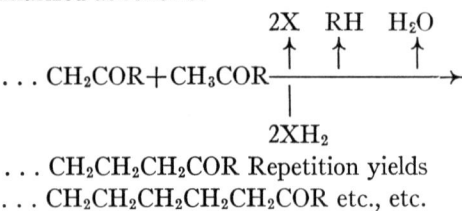

... $CH_2CH_2CH_2COR$ Repetition yields
... $CH_2CH_2CH_2CH_2CH_2COR$ etc., etc.

Another reaction which involves taking up hydrogen as the sustaining factor is the formation of certain amino acids from non-

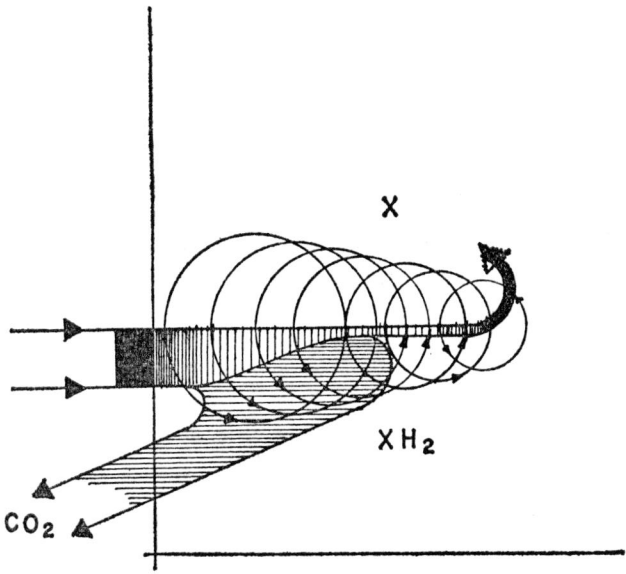

FIG. 5.—Schematic diagram of the substrate conversion in all living cells. The black areas represent complicated carbon material; the hatched ones, simplified material. The degree of shading thus represents the degree of complexity. We see (*to the left*) how complicated substrate—nutriments of different kinds—wander into the cellular machinery under continual simplification, which we can call a normalization process. The main part of the simplified wanders out again in the form of the very simplest: carbon dioxide and water. The series of circles symbolizes the reactions that split off hydrogen from the normalized substrate, thereby making the material more labile so that it can gradually split off carbon dioxide. The hydrogen-charged acceptor substances, here symbolized as XH_2, react with a small fraction of the simplified material in order to build up, step by step, something more complicated within the cell framework, i.e., cellular material, which we see gradually becoming blacker to the right in the figure; this is the beginning of a storage of the complex in all variants—fats, carbohydrates, and proteins.

nitrogenous carbon compounds, in which nitrogen, as ammonia, is incorporated according to the following:

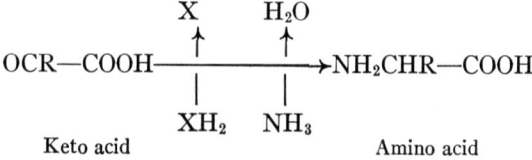

$$\text{OCR—COOH} \xrightarrow[\text{XH}_2 \quad \text{NH}_3]{\text{X} \quad \text{H}_2\text{O}} \text{NH}_2\text{CHR—COOH}$$

Keto acid   Amino acid

The diagram is naturally extremely simplified.

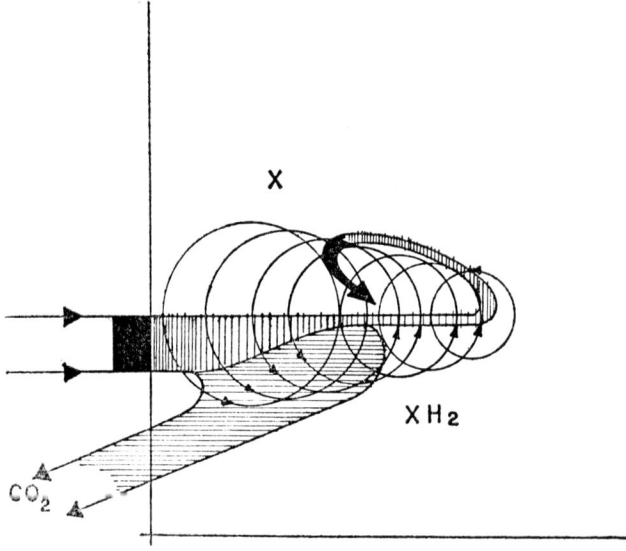

Fig. 6.—The diagram in Figure 5 is supplemented here with a recoupling swing, which suggests that the complex material formed within the cell is not something static, inert, but is subject to a continual simplification via the splitting-off of hydrogen and liberation of carbon dioxide, in the same manner as the main part of the entering substrate. We can interpret this internal decomposition of cellular material as metabolism of endogenous substrate, a parallel to the digestion of nutriments, the decomposition of exogenous substrate. The mechanisms are largely the same. The figure suggests, in other words, that the cell is something dynamic—a transitional stage for a flux of material under continual decomposition and synthesis.

Likewise for the carbohydrates, synthesis from simpler material must involve one or two steps of hydrogenation. We can make a general observation here that each cell in its digestion of nutriments is faced with a situation where the excess of hydrogen, XH_2, can be somewhat of a problem. The individual solutions of the problem lie in the distribution of labile hydrogen. In certain cases the excess hydrogen is utilized mainly to form fatty acids; in other cases the emphasis is on carbohydrate synthesis; and in others amino acid production is the dominating activity, in and for later synthesis of proteins.

In all this, there is a concomitant leakage of hydrogenated products, such as lactic acid or alcohol, depending on the nature of the cells and the environment in which they exist. Lactic acid or alcohol is a rather common end-product from cells growing in an oxygen-deficient environment. In ordinary oxygenous surroundings, the chief end-product is simply water, H_2O. From this point of view, what we call *growth* may be regarded as the automatic result of a situation where the removal of bound hydrogen in the form of water and other leakage products functions so ineffectively that the excess in certain cases *must* be stored within the cell as fat, carbohydrate, and protein—i.e., as cellular material.

As a thesis, this is a somewhat venturesome interpretation without more precise evidence. However, in order to obtain a better grasp of the subject than this superficial presentation of the growth problem can give, we must digress to the region of energy metabolism, and from there to protein synthesis and what controls the formation and activity of enzymes. After returning from this probably rather laborious excursion, we can then consider what we perhaps comprehend only as a paradox, but which in reality comprises much more than we can surmise.

To plunge directly into the energy metabolism, we observe first that hydrogenated carbon material of cellular substance has higher energy content and a higher degree of complexity than the simpler carbon compounds from which it is synthesized. Hydrogenation per se can in many cases raise simpler compounds with low energy content to a higher energy level; but in order to produce at the same time the more complex from the less, simpler units must often be *activated*, which process is accomplished in cells of all kinds with the help of a series of *labile phosphate compounds*.

In themselves, phosphates on the inorganic level, represented by phosphoric acid, H_3PO_4, and its salts, are fairly simple units without any external chemical characteristics that make them particularly remarkable. One peculiarity is the tendency of certain phosphates to couple several units to a greater aggregate with the formation of water. Monophosphates can be coupled to di- and tri- up to polyphosphates with energy added on the inorganic level in the form of heat.

In the living cell there is a similar but more complicated coupling of phosphate units. In the simplest form of the mechanism, certain intermediate products of metabolism, during the course of simplification and destruction, are involved in a dehydrogenation reaction in which inorganic phosphate also happens to be present. An example is the dehydrogenation of certain compounds in the direction of the formation of corresponding acids of carboxylic type:

$$R-CHO \xrightarrow[XH_2]{\overset{X \quad\quad Phosphate}{|\quad\quad\quad |}} R-CO\text{-phosphate}$$

a parallel to the formation of the corresponding acid

$$R-CHO \xrightarrow[XH_2]{\overset{X \quad\quad H_2O}{|\quad\quad\quad |}} R-COOH$$

As for their chemical nature, these organic phosphates, R—CO-phosphate, or more correctly, $R-CO-OPO(OH)_2$, are relatively reactive. *The dehydrogenation reaction has thus yielded as a by-product a compound with chemical aggressiveness*, ready for a great many different transformations. A particularly important one is the coupling with another organic phosphate, leading to the formation of organic *triphosphates*. Bluntly expressed:

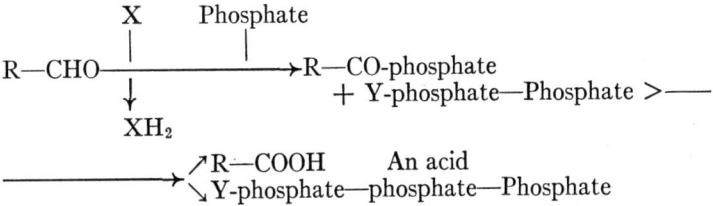

We see that just as X can act as hydrogen transfer unit in the form XH_2, what we have called here Y-phosphate—phosphate acts, in the same manner, as phosphate transfer unit in the form Y-phosphate—phosphate—phosphate. The technical names for the latter compounds are adenosine diphosphate and adenosine triphosphate, abbreviated ADP and ATP. (Parenthetically it may be said that there is a whole series of similar and closely related phosphate transfer units.) In summarizing the interaction between hydrogen transfer and phosphate transfer, we can condense the above reactions to:

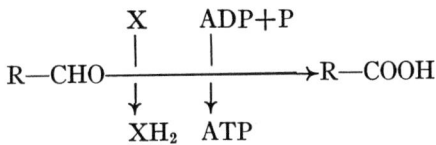

P stands for inorganic phosphate.

Here is a fundamental reaction type which is a universal feature of all living cells. This implies the following. In certain dehydrogenation reactions, hydrogen transfer via X—XH_2 is accompanied by an incorporation of inorganic phosphate. The reactive phosphate compound formed as the intermediate stage can, via ADP, donate its phosphate to ATP, which now represents *stored chemical reactivity. Thus, at the same time, we have a storage of hydrogen as XH_2 and of chemical aggressiveness as ATP.* The combination is of decisive importance for all synthesis within the cell, because stored ATP can in turn transfer its charge of stored reactivity to other inactive compounds and thereby activate them to new combinations. An attempt to symbolize this is shown in Figure 7.

The formation of ATP discussed here takes place in all cells, regardless of the environment, even where oxygen is lacking. Thus, it is a situation that is common to a great many lower organisms, from bacteria up to the tapeworm. The yield of ATP from cells in an anaerobic, oxygen-free environment is not particularly large, since only a few dehydrogenation reactions involve simultaneous phosphate storage. One molecule of glucose digested by an anaerobic organism yields during its enzymatic decomposition only two mole-

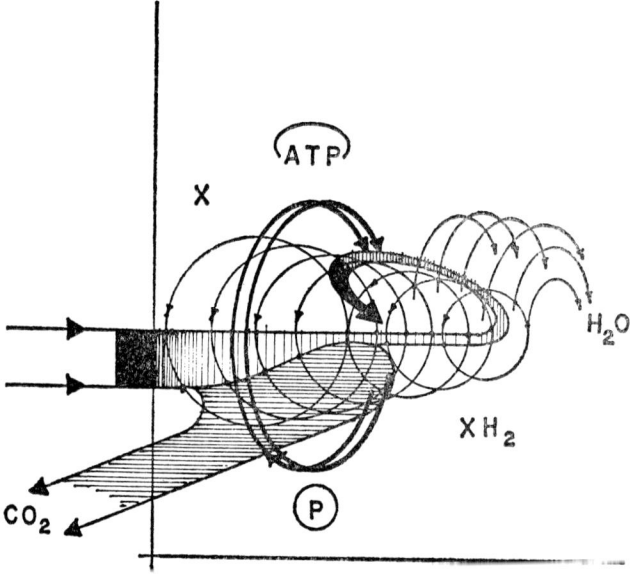

FIG. 7.—In this figure two vertically running cyclic processes have been inserted, symbolizing one phase of the energy metabolism in cellular metabolism. In certain steps of the splitting-off of hydrogen from exogenous and endogenous substrate, energy is gained in the form of labile and reactive phosphate, here symbolized as ATP. This stored energy takes part in the resynthesis of cellular material together with hydrogenation reactions, where XH_2 gives up hydrogen to simple material under re-formation of acceptors of X type, ready for a new round. The arrows curving to the right denote that water is given off in these dehydrogenation reactions. The whole figure thus symbolizes a living cell under anaerobic conditions, where the internal yield of ATP, and thereby of newly formed cellular material, is quite insignificant in relation to the amount of nutriments that have been digested.

cules of ATP as a by-product in the dehydrogenation reactions. In aerobic, oxygenous surroundings, on the other hand, there is a remarkable increase in the yield of ATP; one molecule of glucose, gradually decomposed with carbon dioxide as the end-product, yields no less than 34 ATP.

Since the combination of XH_2 and ATP is the essential factor for the activation and transformation of simpler carbon compounds into cellular material, there is thus a considerably greater chance of growth activity for aerobic organisms than for anaerobic. Or, more correctly, the aerobic can utilize nutriments considerably more effectively than the anaerobic. The reason lies partly in the special type of ATP production by the aerobic cells, which, in addition to the reactions we have discussed here, can also form ATP via the fundamental reaction:

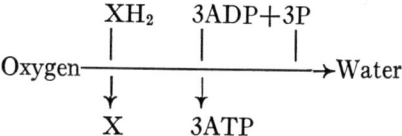

For each XH_2 that donates its hydrogen to *oxygen*, forming $X + H_2O$, an average of 3 ATP are formed as a by-product. This extraordinarily effective energy storage in the cell reflects the framework of a number of complex enzymatic organizations we call *mitochondria*. Within these busily dehydrogenating enzyme complexes the complete decomposition of nutriments to carbon dioxide is brought about. Each available unit of XH_2 is confronted with oxygen, ADP, and inorganic phosphate in intricate interaction, which leads to the production of ATP, as symbolized in the above formula —namely, what we call *oxidative phosphorylation*. See Figure 8.

We can pause here to catch our breath and, by breathing oxygen, let the mitochondria in our cells build up an extra supply of ATP. (We probably need it at this stage, after our deep plunge into the chemical labyrinths of cellular metabolism.) As a diversion we can glance at Figure 9, which is an ambitious attempt to symbolize *essentia chemica* for a photosynthesizing cell as a whole.

The tendency to activate the chemically inert via ATP and the action of organic phosphate compounds appears again in an extreme form in what we call photosynthesis. We have come in contact earlier with the concept summarized in a verbal form: that green plants consume carbon dioxide and water under the activating influence of sunlight. On the level which is represented by cellular metabolism we can conceive of the process of photosynthesis as something chemically related to the activity in aerobic cells, as it is presented in Figure 8.

The abstract representation shown in Figure 9 is an attempt to

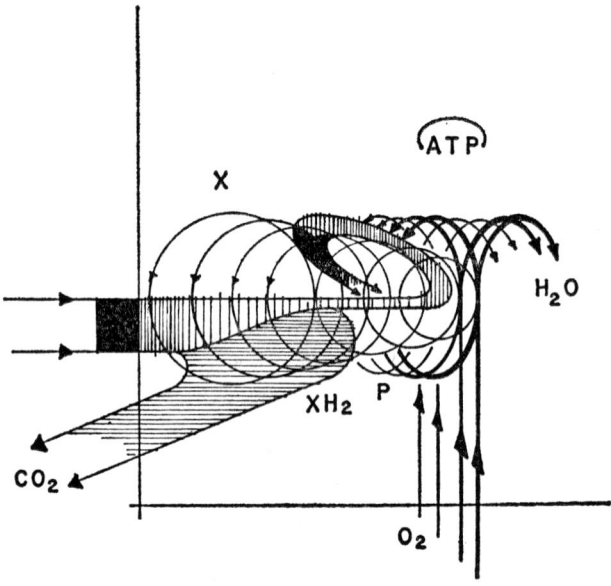

FIG. 8.—Here the conditions shown in Figure 7 have been markedly effectuated by the introduction of oxidative phosphorylation, which is shown at the right as vertically running cyclic processes coupled to some of the circles symbolizing hydrogenation-dehydrogenation, i.e., XH_2—X—XH_2. . . . The ATP that is gained here—in a remarkably effective yield—has a direct connection with the oxidation of some XH_2 through the action of oxygen. For each XH_2 that in stepwise reactions is confronted with oxygen and yields $X + H_2O$, 3 ATP on the average are formed. This effectively formed ATP supplies energy for the new synthesis of cellular material—growth processes in general, here symbolized as an increasingly shaded curved region of internal material. An active cell under aerobic conditions, with effective energy metabolism.

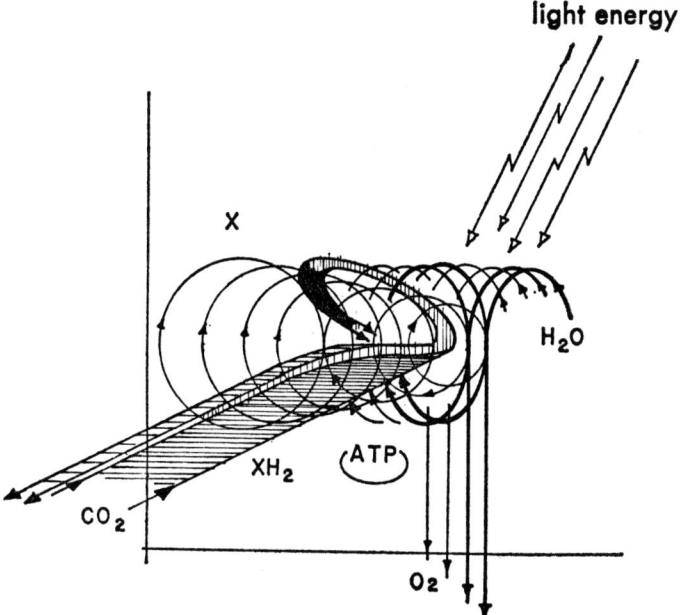

Fig. 9.—This figure attempts to symbolize the conditions within a cell capable of photosynthesis. The substrate here is carbon dioxide, which in the lower left corner is seen to wander into the system. If we compare this figure with the four preceding ones, we see that the hydrogenation-dehydrogenation circles here run in the opposite direction, symbolizing a predominant tendency to hydrogenation of the simple material that is first formed from carbon dioxide and carbon dioxide acceptors. The perpetual new formation of hydrogen-charged acceptors, XH_2, takes place here in the centers where water under the influence of radiant energy is split up into oxygen and hydrogen, the latter being attracted to X. A certain amount of ATP is also formed in connection with the water cleavage, which in combination with hydrogenation processes of all kinds converts the simple carboniferous material, which is formed here from carbon dioxide, into increasingly more complicated cellular material. The synthesizing reactions of the plant cell have, however, a certain antipode in some reactions where carbon dioxide is again split off and released from the system—which is drawn in the left corner as a smaller stream from the system. Respiratory processes of the plant cell leading to the maintenance of a certain dynamism. See also Figures 10 and 11.

illustrate the principle of the mechanism. For a more concrete picture of the whole process, the reader is referred to Figure 10 and the more technical representation in Figure 11.

As we see from these summarizing diagrams, especially from Figure 11, carbon dioxide enters into a situation where a highly phosphorylated compound, here called RP, forms a product with it,

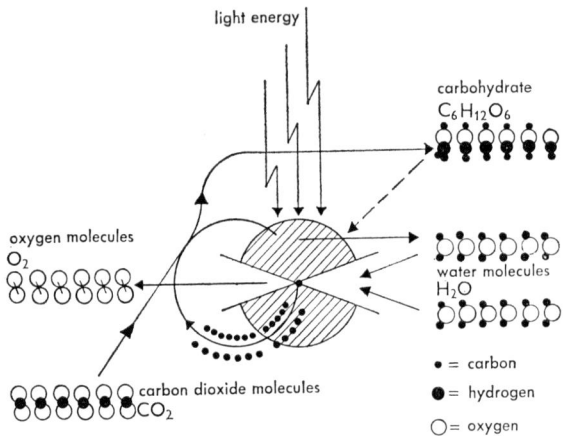

Fig. 10.—A crude attempt to illustrate the phase in photochemical activity of green plant cells where water is split up into hydrogen and oxygen, and carbon dioxide is gradually assimilated, forming as the first storage product carbohydrates of different types, primarily simple sugars and then more complicated compounds, e.g., starch.

$RP.CO_2$. At a later stage this product, upon hydrogenation, is converted into something that we can symbolize as R′P-CHO, the stage from which carbohydrates and their conversion products can be formed. The driving force of the whole process is solar energy acting on green plant cells in which organizations of special enzymes + chlorophyll, chloroplasts, split up water in the presence of hydrogen acceptors, yielding XH_2 and oxygen, as ATP is simultaneously formed. The latter transfers its chemical aggressiveness to a compound R, which in the form RP is adapted for accepting carbon dioxide. The addition product $RP.CO_2$ is stabilized by hydrogenation to R'P.CHO ..., from which are gradually formed carbohydrates, cellular material, and more R, which in turn can take up more ATP, forming RP, which can accept CO_2, forming RP.CO_2 ..., and so forth.

Figures 9, 10, and 11 are attempts to represent in different ways

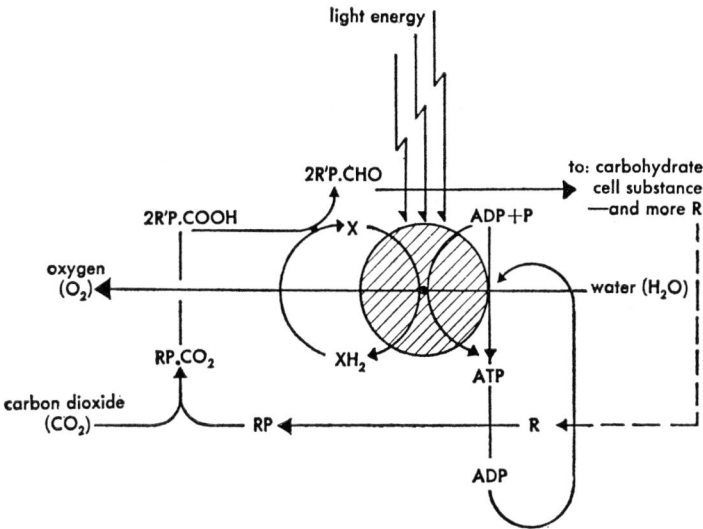

FIG. 11.—Diagram of the simplest detail pattern of photosynthesis, showing that the photochemical formation of ATP and the cleavage of water into XH_2 and oxygen are most intimately connected to each other within the framework of the chloroplast system of the green plant cells. Chloroplasts are chlorophyllous organelles within the cells, where what is symbolized above by a hatched circle takes place.

the chemical logic of photosynthesis. Figure 9 has a small detail that is worthy of a separate discussion. A wide path has been drawn for the *intake* of carbon dioxide and a narrow one for its *output* to symbolize the fact that plant cells breathe, even the photosynthetically active ones. One can estimate that, on the average, 15 to 30 per cent of the carbon dioxide taken up by plants throughout the world is released again from the plant material as a whole through dehydrogenation processes similarly to what takes place in animals in general. Only 70 to 85 per cent of the absorbed carbon dioxide is converted into the cellular material of the plants, not more, but that is quite sufficient for synthesis to outweigh decomposition. In the many cell types of the animal world we have in many respects a reverse situation, in that certain reactions in the cells involve a temporary incorporation of carbon dioxide, but its liberation as a consequence of dehydrogenation clearly outweighs the processes leading to incorporation.

In our efforts to grasp something of what occurs or seems to occur, we have landed in a world where color and form lack meaning, and where function can be expressed only through a network of chemical symbols. We have grasped something of the cell in its destructive activity, its continual simplification of the structure of the nutriments, to yield hydrogen and stored energy for new combinations, for synthesis. In order to get an idea of the whole, what we saw as a cell in totality, we must now work up gradually toward a reconstructive view of the chemical network in activity, both breaking down and synthesizing. We have quite an arduous task ahead of us. Have we stored up enough ATP to keep us going?

We can make a nice start by pondering over what we have hitherto ignored for the sake of simplicity; namely, the nature of protein substances and enzymes. We noted in passing that all proteins are composed of a number of units, amino acids, which in a chainlike pattern are built into highly complex architecture. The pattern varies for each type of protein and thus for each type of enzyme. The abundant variation we see in nature is derived to a certain extent from the enormous variability with which twenty-odd units can theoretically be combined within the framework of structures comprising up to 10,000 or more units! One of the main problems of biochemistry is just this: how are the twenty different simple amino acid units, all of the general structure NH_2-CHR-COOH, combined into *exact architecture* within the structural pattern of the giant molecules?

The preliminary steps, at least, appear to be understandable in principle. The first phase of protein synthesis starts with each amino acid, which is a quite inert unit in itself, being activated by ATP into a reactive combination which chemically links the amino acid via its -CO group to the phosphate. Compounds of this type have great potentialities for forming long chain structures through self-condensation, somewhat in the following manner:

$$\ldots NH_2CHR-COOH \quad NH_2CHR-COOH \quad NH_2CHR-COOH \ldots$$
Amino acids+ATP   +ATP   +ATP

$$\ldots NH_2CHR-CO \quad NH_2CHR-CO \quad NH_2CHR-CO \ldots$$
$$\setminus AMP \quad \setminus AMP \quad \setminus AMP$$
(+PP)

$$\ldots NH_2CHR-CO-NHCHR-CO-NHCHR-CO \ldots$$
(+AMP)

A longer *peptide*, which will be coiled up in a special way to form a part of a *protein*.

Thus a number of activated amino acid units in combination with AMP confronting one another could result in peptide chains. So much is obvious. But how does this confrontation take place, and what controls the process so that each amino acid lands in the right place? If we take a look at Figure 12, which is a representation of an extremely simple protein substance, insulin, it seems almost inconceivable that any process could build up such precise architecture. And yet every minute each cell builds thousands of such giant molecules, with varying structural pattern for each type strictly repeated from case to case. Organization out of chaos; but what does the organizing?

During the past twenty years nucleic acids, giant molecules of entirely different structure, have been suspected of organizational activity within the cell. A nucleic acid is a chain structure, the constituents of which have a definite relationship with what is called for the sake of brevity ADP and ATP. One type of nucleic acid, designated RNA, ribonucleic acid, contains four different types of units coupled in sequence by phosphate linkages; the *patterns of the sequence* constitute codes which have particular effects on protein

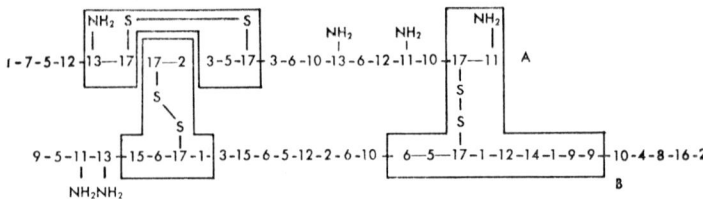

FIG. 12.—An exceedingly simplified diagram of the composition and structure of a simple protein substance, in this care insulin. Each number denotes a certain definite structural unit, an incorporated amino acid residue, one of the units $-NH_2-CHR-CO-$ in the chain pattern, where each number symbolizes one of the 17 variations of chemical structures. Each unit, in the form of the corresponding amino acid, $NH_2-CHR-COOH$, has its special name, as shown in the table below. A more detailed description of the structure of all these units is not possible here due to lack of space. The interested reader is referred to textbooks of biochemistry for this information. We can mention here that of the units, no. 17, systine, contains one sulphur atom, S, at the end of the group, which makes possible the combination of the two peptide chains A and B into a larger unit by formation of bonds of the type S–S. Thus we actually have four "half-cystine" units in the A chain and two corresponding ones in the B chain. Each protein is an individual variant of this theme of peptide chains of different types, linked by –S–S bonds and other coupling elements into units of often extremely intricate architecture. The number of units—here 51—can often amount to many hundred; in some cases to over a thousand. The structural analysis of proteins is an exceedingly difficult undertaking.

EXPLANATION OF NUMBERS

1. Glycine
2. Alanine
3. Serine
4. Threonine
5. Valine
6. Leucine
7. Isoleucine
8. Proline
9. Phenylalanine
10. Tyrosine
11. Asparagine
12. Glutamic acid
13. Glutamine
14. Arginine
15. Histidine
16. Lysine
17. Cystine

synthesis. Figure 13 shows the principle of the structural pattern of RNA in schematic form.

The idea presently in vogue is that each type of protein synthesis is *governed* by the presence of a certain type of RNA, acting as an organizational unit to create order in reactions which would otherwise theoretically represent *all* mathematical possibilities of aminoacid condensation.

Therefore, as a first summation: *every synthesis of a specific protein is organized by nucleic acid, type RNA, in one of its structural*

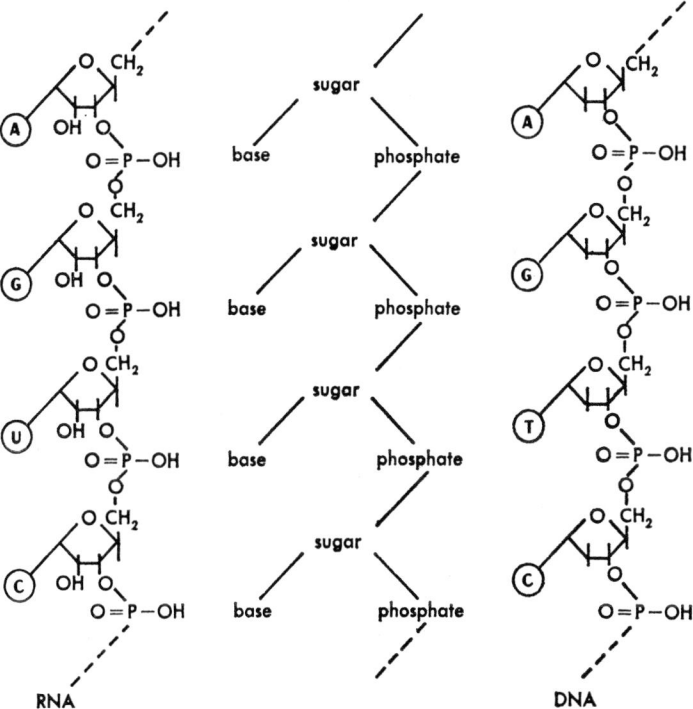

Fig. 13.—Exceedingly simplified schematic presentation of the structure of nucleic acids. The bases A, G, U, C, and T go under the technical names adenine, guanine, uracil, cytidine, and thymine. What is designated here as sugar is a pentose, called ribose (RNA) and deoxyribose (DNA). Phosphate units link the molecule together into a giant formation.

forms, which themselves are determined by the code-like repetition pattern of four units.

It is still somewhat of an unsolved mystery how the RNA structures with their code of four constituent units can conceivably control the occurrence of highly specific protein structures composed of some twenty structural elements. Furthermore, the question is, can we accept the idea that the synthesis of every protein is governed by a definite RNA structure? There is a slight chance that each RNA in combination with a number of activating amino-acid units could lead to *a certain sequence* of amino-acid units, and that this sequence could reappear in many proteins; in other words, every protein synthesis would not be strictly bound to a certain type of RNA but to an interaction of several RNA, each governing a certain sequence of amino acids. Such a case would be an expression of a hierarchical organization, with a smaller number of types of RNA than types of protein. The top rank in such a hierarchy would be a superorganization, the *genes* of the cell nucleus, which contain another type of nucleic acid, type DNA, deoxyribonucleic acid. DNA is known to govern the formation of specific types of RNA, which in turn organize a greater number of specific protein syntheses.

We shall return to this difficult problem later. Here we can say that the formation of proteins, and thus of enzymes of all kinds, has its organizational basis in the genes of the cell nucleus, within the chromosomal organization that generation after generation exhibit structural stability; at the same time, the interaction of cells on the level of sexual differentiation creates continuity in the patterns of variation. In cells of higher organisms all this appears as a chain of events with feed-back influence: the chromosome structure of the cell nucleus, with genes in different loci; the DNA that influences the formation of RNA in its different structures; RNA in turn reacting on protein synthesis, the products of which are the structural elements of the cell and its enzymes; enzymes engaged in metabolism of nutriments as well as in *synthesis of the different structural elements* of RNA and, moreover, involved to a certain extent in the synthesis of DNA. And the cycle continues.

It is still an open question how the enzymes operate jointly in the network of step-by-step processes which together constitute the individual patterns of cellular metabolism. In certain cases, such as the intracellular structural elements we call mitochondria, for instance, we have definite evidence that some twenty enzymes within one and the same limited area successively digest certain metabolites according to some kind of cyclic process. This is shown below; it is a variant of the general scheme we often set up to symbolize enzymatic reactions in coupled sequences:

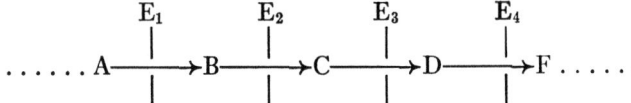

Behind the existence of every enzyme lies an organization of genes that determines the structure and amount of each participating enzyme in the chain, one of the many in the network. But if we should examine more closely the functioning of such an enzymatic process —where substance A is converted step-by-step by means of the enzymes E_1, E_2, E_3, E_4, via B, C, and D, into F—we notice that our symbolism has idealized the conditions to an alarming degree. For a clear and frictionless conversion of A to F, it is essential that all steps function equally effectively. Do they do this? Let us assume that a certain cell type has such a finely co-ordinated set of enzymes that they can convert A into F with a yield of 100 per cent. This would represent the ideal catalyzer, self-contained and unchangeable. On the other hand, we know that all fluctuations in external environment have an influence on the effectiveness of enzymatic reactions such that an acceleration or retardation at one point must affect the whole system. In the chain A–F above, only a slight inhibition of E_3 would produce a cumulative storage of product C. If this tendency continues, C slowly increases within the cell so that certain secondary consequences become noticeable. There are other possibilities: C can simply leak out into the surrounding medium— we observe this as a simple fact and say that this cell type *produces* this and that, in this case product C. If C within the cell should come within the reach of a number of enzymes that should form a sparingly soluble compound out of C, naturally the latter would remain

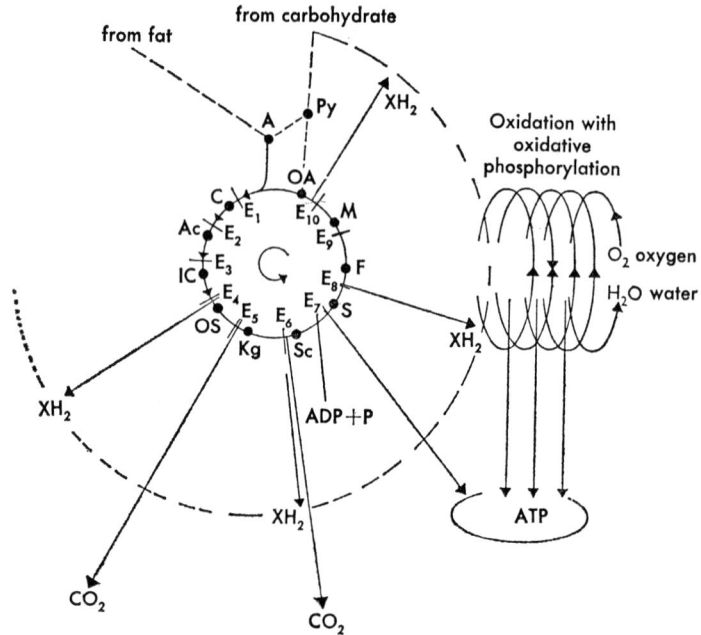

FIG. 14.—Symbolic representation of the central mechanism that in the cell liberates energy through terminal oxidation of acetic acid in activated form. This active acetic acid, here designated A, is the final stage for the intermediate decomposition of both carbohydrates and fats and certain amino acids. The diagram shows the acetic acid A combining with a substance OA, forming a substance C. Further transformations of C via many steps lead to the reformation of OA, ready for a new round with new acetic acid. Each step is catalyzed by a special enzyme, one of these designated here as E_1, E_2, E_3, ... In mitochondria of higher cells the interaction of these enzymes has reached an extraordinarily high degree of organization. In the process two molecules of carbon dioxide are split off, just the number of atoms that correspond to acetic acid, CH_3COOH. Also, four XH_2 and one ATP are liberated from the cyclic process. The confrontation between four XH_2 and oxygen in the enzymatic system of the oxidative phosphorylation can give a maximum yield of 12 ATP per round.

ABBREVIATIONS

A	Activated acetic acid, acetyl-CoA	Sc	Succinyl-CoA
C	Citric acid	S	Succinic acid
AC	Aconitic acid	F	Fumaric acid
IC	Isocitric acid	M	Malic acid
OS	Oxalosuccinic acid	OA	Oxaloacetic acid
Kg	Ketoglutaric acid	Py	Pyruvic acid

within the cell. In certain cases, moreover, C, as such or transformed, *can affect* one or another enzymatic reaction in the network; for example, a series:

$$\ldots\ldots X \xrightarrow{E_{11}} Y \xrightarrow{E_{12}} Z \xrightarrow{E_{13}} K.$$

An inhibition of E_{12}—not much, only a little—would lead to a slow accumulation of Y, which either leaves the cell or is stored or undergoes metabolism through other channels. If we pursue this line of thought, we find that the chances of complete conversion of nutriments to only soluble leakage products are relatively small. Sooner or later something must be accumulated within the cell. And we know that harmonious accumulation of all cellular constituents is growth. Do we see a connection?

Let us reflect for a moment on protein synthesis: nutriments converted to keto acids, which with XH_2 and nitrogen compounds become amino acids. These are so activated by ATP that in the presence of RNA they cannot avoid becoming proteins. Some of these proteins are of the type that encourage the conversion of the substrate into more amino acids. Some encourage the formation of more RNA, thus more proteins, more enzymes, more cellular material—growth! Gradually the process leads either to cell division or to death and dissolution through unharmonious nutrition. What we see as the all-pervading principle, however, is that cellular growth processes are coupled to a network of enzymatic reactions in such a way that growth in most cases *is inevitable*—provided that the environment contains the nutriments for this process; or, in other words, the force of circumstances brings cells together with their special substrate!

Do we recall having heard this earlier? Yes, in the very beginning of this chapter, where we saw under the microscope a cell in detail, swimming in a nutrient solution, engaged in budding, growth, division, while outside the environment someone for some reason was confronting the cells with their special nourishment and thereby consciously starting the automatic reaction which for the majority of cells is *growth*, division, formation of new cells of the same type, new generations—all giving in its external manifestations, an illusion of purposeful expediency. At the same time, however, we have gradually found out that the coupling pattern for all enzymatic reactions of the cell—that is, cellular function in its entirety—is so formed that each alteration in the composition of the milieu induces an automatic readjustment of the system in the form of subtle changes in one or more enzymatic reactions. This readjustment—which in turn influences the system as a whole—is included as a matter of principle in what we call *adaptation*, a process which often involves a slow but thorough reorganization of certain enzymatic syntheses with special consequences for metabolism as a whole. Often the final result of adaptation is almost unnoticeable so far as the external behavior pattern of the cell, so to speak, is concerned. At times, however, we observe one or two striking chemical consequences; for example, the production of some substance as the result of a certain inhibition of some enzymatic reaction. This can result in an intracellular accumulation of the substance or, in other cases, in a leakage out into the environment, *which is thereby altered*, with consequences for other cell types which may happen to be there. The entire system of coupled enzymatic reactions in a cell has in this way a feature of *chemical alertness*, of sensibility to environmental fluctuations. According to this viewpoint, every external alteration results in an intracellular response through automatic adaptation to the new situation, most often in the form of readjustment of the enzymatically engineered production pattern; in certain cases in the extremely minute increase of some cellular material, resulting in growth and cell division. In certain situations of abrupt environmental change, the capacity for readjustment is exceeded; the result then is disorganization, dissolution, death. All these manifestations of chemical alertness are something we can observe when we keep a population of simple cells under control in an environment of such a composition that the existence of the cells is assured for a time. Such is the situation when the experimenter consciously alters external conditions; for example, as part of a research program where the optimal milieu for growth is analyzed and where blocking agents are introduced for studies of the production pattern or with the intent of injuring and killing. The cells are *exposed* to environmental fluctuations and their automatic response is formed individually according to their genetically determined enzymatic pattern.

The same situation exists, although in another form, for all cells comprising a more or less complex organization. In every more or less complicated organism the local environment for certain cells is the combined result of the production pattern of all cells, influenced in turn by the organism's ingestion of nutriments as they are determined by its own external environment. What we observed as chemical alertness and adaptability on the cellular level reappears here again in summation in the general behavior pattern of the organism. We see again, in the organism as a whole, something of inevitability in the effect of environmental pressures, which in complex cellular organizations trigger finely adjusted automatisms in a certain chemical direction, counteracting all kinds of fluctuations; what we call *homeostasis*. All this results in complex external manifestations, such as behavior pattern, adaptation, and automatic reactions that preserve the individual existence—for a short time. The external environment here is other organisms in the same situation, in growth, activity, and dissolution. The driving force of the whole process is the energy radiated from a distant sun.

We have attempted, starting from the study of a cell seen in detail, to obtain a glimpse of something else that we were seeking but perhaps did not find. The cell under the microscope, in its external reactions, its external structure, was undoubtedly something living. When we began to dissect the chemical process behind the morphologic façade, we found the complex network of metabolic reactions, the degradation and complication of carbon compounds within a limited structure. Furthermore, we found a kind of chemical conservatism in the stabilizing influence of the genes on the whole, enzymatic repetition according to a given individual pattern, while at at the same time a flexibility, which we called chemical alertness in the form of perfected automatisms. Finally, there was the capability of self-propagation, maintenance of the typical pattern throughout a series of generations.

In all this we have glimpsed something of individuality and sensibility, something which we could find again on the organism level in another form, which we have within ourselves, and which we sometimes believe is other than what it really is. What seems to be so difficult to grasp in all analyses of cells and organisms is just this contrast between the manifestations of individuality and sensibility that we can observe in all cellular organizations and, at the same time, their automatic determination by the environment; what takes place as a network of individual reactions in cells, cell groups, and organisms is strictly connected to the great whole, to the activity of all cells, now, and with direct connection to their developmental history—in other words, to life in general. All this is an expression of the two aspects of life that we must accept. Earlier we saw the system as a whole. Here we have seen a cell in detail, and thus have obtained a kind of least-common-denominator for the whole. Did we get the answer we were looking for?

CHAPTER IV
OVERVIEW

The development of the potentialities of matter, which here on our planet has resulted in organisms of all kinds, has with the emergence of man introduced a biologic innovation for the region: a highly specialized form of time-consciousness. It is a property that we human beings of today begin to comprehend vaguely as a variant of that sensitivity to fluctuations of all kinds which characterizes cells and organisms in general. We begin to be able to distinguish how this sensitivity to intervals, this chemical alertness, has in certain organisms been carried so far that transitory episodes can be recalled; how instinct has gradually developed into memory and consciousness; how in man the difference between *past* and *present* has begun to be realized. Step by step our interval of consciousness has widened to remembrance of occurrences of several days ago. The concepts of "yesterday," "many days ago," "many solar cycles ago," have begun to take form; and finally the idea of a future has come to be established in our primitive brain. The span has been rapidly widened; using the solar cycle, the year, as unit, we register the main features in the developmental history of our planet and people, and we learn to comprehend our own situation in the present, between the past and future. As an inevitable consequence of our development into conscious living organisms, we have gradually begun to analyze the necessary conditions for life both on our planet and in other localities in the universe—this process that involves a transformation of unconscious matter into such a degree of complexity that the material can under certain circumstances begin to analyze its own origin.

So far as our temporal perspective is concerned, we have gradually become conscious of the fact that our own evolution—to *Homo sapiens* from advanced primates—represents a minimal fraction of the time that has passed since the formation of the earth from cosmic dust. If we let this span of time symbolize one year, we find that from January 1 to December 1 there is little to be noted concerning organisms as we define them. By the middle of December, however, we can ascertain that life exists in the form of quite advanced organisms in the Cambrian seas. The latter part of the last week of the year finds dinosaurs in full activity and the prehistoric bird Archaeopteryx flying with difficulty between primitive pine trees. During the last two days of the year some newcomers appear on the scene, some mammals, the saber-toothed tiger representing their uninhibited vitality. During the last day, at 2300 hours the first manlike creatures appear: Sinanthropus, Pithecanthropus, and their African colleagues. The last minutes show the drawings in the Altamira cavern and the last glacial epoch. The last second of the year contains all the latest news: the fall of the Bastille, the battle of Trafalgar, Darwin's *Origin of Species*, two world wars, and the development of the vacuum cleaner. In the last tenth of a second come the utilization of atomic energy and fumbling attempts to reconstruct the origin of life.

This is the time scale, the last moment of which contains the hectic development of natural science and technology during the past few years—an exceedingly small fraction of the total time span. As a link in our attempts to understand something of the structure and forces of the world around us, the question of the development of life on our planet has become the object of some interest as a problem with certain stimulating aspects. The problem itself does not constitute one of the foremost questions of natural science; actually there are more immediate problems just as stimulatingly complex. However, the special flavor of adventure in the problem of the origin of life probably stems from its controversialness, for a remarkable lack of certain data affords opportunity for highly indi-

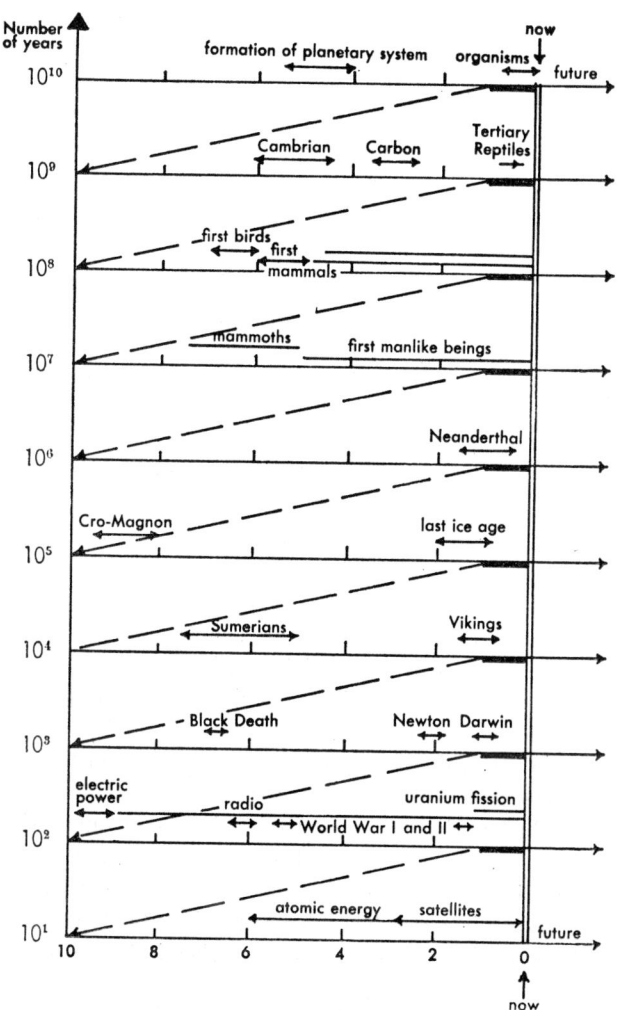

Fig. 15.—An attempt to visualize the development of the earth and organisms—including our own—within a common time scale. The top row represents 10 billion years. Each subsequent unit of the scale from the top to the bottom represents the last tenth of the preceding scalar unit, extended 10 times.

vidualized theorizing. This is true of all problems in their earliest stages. The next generation of natural scientists will certainly have a better foundation to build on than we have now, for we are still uncertain about the actual nature of what we are seeking in our attack on the problem of the origin of life.

In certain respects, the situation requires unusual concentration. In the first place, are we adequately prepared to meet the question: "Is it the origin of organisms in our meaning that is the goal of our quest, or is it something else that is involved in life as a system?" The question has a certain weightiness.

Are we seeking a way to reconstruct a moment of dramatic force, to see before us an accumulation of inorganic material and primitive carbon compounds, the raw material of life, in a limited region, all prepared for a sublime transformation into the first living primordial cell, the descendants of which shall populate the earth? A moment when a balanced combination of the proper ingredients happens to be subjected to the proper pressure, the proper temperature, the appropriate circumstances for the spontaneous formation of a primitive cellular material with the ability to reproduce itself in a world of simple and uncomplicated substrates?

Such an attempt to see the origin of life as something fixed to a particular occurrence puts us in a difficult situation. A simple analysis of the chances for momentary conversion of simpler carbon compounds into something so complex as a cell—even in the simplest form—shows that the probability is practically non-existent. Despite this, the fact that life evidently was formed on our earth in the far-off past must indicate, from this point of view, that something other than the properties of the material was responsible for the origin of all the incomprehensible complexity that later developed into living organisms. It appears to be a chemical miracle initiated by something outside the system.

Two not quite unexpected expedient solutions have been proposed. The first is straightforward and involves the intervention of divine power that carries out in an antistatistic direction the first sublime chemical synthesis of a cell. To a certain degree this is a dramatic conception that goes back to wholehearted efforts of the Middle Ages to force refractory material in the direction of the desired result by means of advanced alchemy spiced by interventions of extraterrestrial nature.

The second proposal avoids the whole difficulty to a certain extent by assuming that the origin of life here on our earth is the result of an infection of spores or virus, or something similar, from another world. The Swedish scientist Svante Arrhenius worked out a plausible theory about fifty years ago on how certain spores, whirled up into the higher layers of the atmosphere of a planet such as ours, should be able to leave the mother planet and, with the help of light-pressure, be widely distributed around in outer space, ready to be swept up by some other planet with suitable nutriments.

This theory per se has some positive points, and the idea itself of a dissemination of virulent particles by means of light-pressure unquestionably provides a fresh grasp on the subject. On the other hand, there are a great many difficulties along the way that hinder a general acceptance of the idea. First, we must realize that a long time in outer space close to sources of cosmic radiation and ordinary ultraviolet light must be exceedingly destructive for spores or viral types, no matter how resistant they may be. Furthermore, they would have to be able to function—i.e., reproduce themselves—in a purely inorganic environment, or in any case in a milieu of chemical compounds of a lower degree of complexity than themselves. Finally, even if we should accept the whole idea as such, where were these spores formed and what in turn was *their* origin? We are in the same situation if we assume that life on our earth was started once upon a time by the landing of a space ship from afar, by conscious implantation of suitable organisms in the future Garden of Eden, or by careless garbage disposal. It is quite possible that we ourselves within the next hundred years shall be able to select suitable organisms to live in the inhospitable milieu of the moon or Venus, and rapidly distribute them there; but what does that matter? The problem of the origin of life has not been illuminated—only the problem of its dissemination.

The more we think about the consequences of life on our earth starting at a certain point, at a certain moment, the more we begin to realize that we must resort to help from the outside, so to speak, to get anywhere, which means that the problem is no longer one problem. We simply put the question completely aside, accept the consequences of a cosmic intervention or an infection from goodness knows where, and with that settle down in peace. At the same time we have a definite feeling that the obstacle to the formulation of the problem is that we unconsciously introduce into the whole an idea of a *beginning*, a starting point, a lifting of the curtain.

How would it be if we, only as an experiment, ventured upon an utterly opposite conception—that life has *always* existed on our earth, ever since the formation of the planet about five billion years ago? At first we might dismiss the whole idea. We can, as a matter of fact, postulate with a certain probability that the surface of the earth five billion years ago was composed of mainly inorganic compounds of a lower degree of complexity plus some 2-, perhaps 3-, atom carbon compounds of relatively simple structure. Three or four billion years later we suddenly find fully developed organisms, floating in the seas, slowly crawling in the lagoons along the shore, searching, irritable, pulsating with life. Is the contrast too great? Are we perhaps so bound by our conception of life as something connected only with organisms in motion, in activity, that we cannot grasp another aspect, that the manifold activities of the organisms are a reflex reaction to something that has existed since the beginning of time, *something that gives us a common denominator for planetary material of a couple of billion years ago and the same material in our own time?*

If we pursue this idea, we find that it is not at all some kind of metaphysical speculation but instead a very realistic starting point for further analysis. This is what we have had in mind, perhaps unconsciously, since we began to observe life both from a distance as a system and in detail as living cells. The path that leads to the origin of cells in our meaning, via something involving a moment in their development, seems to be closed. The other path is open for analysis. It involves reconstructing the transformations of the earliest carbon material between separate loci of destruction and recombination, later leading to a flux of more complicated material between chemically opposite poles, from which something is *gradually* formed which *gradually* becomes what we today recognize as cells

—in other words, an analysis that is based on the postulate that *life is older than organisms.*

We can formulate a working hypothesis something like this: The system we see today of carbon compounds in constant turnover between the depots of inorganic carbon dioxide, plant life, animal life, and back again existed much earlier in another form, long before organisms entered the picture. Further, the opposite poles in the process were other than they are now; the *process* is the primary factor. The development and the structure of the participating units are secondary, something that evolved during the course of the process and led gradually to a flux of participating units, where we recognize cells in our meaning.

An investigation along these lines has some possibilities worth utilizing. We can start with a discussion of life and organisms in the Cambrian period, about 500 million years before our own time. Further back, we shall see how far we can push an analysis of conditions at the more remote time when organisms actually existed, long before calcareous shells and chitin integuments became the morphologic characteristics that we now recognize in fossils from the Cambrian. Then we can start a reconstruction of the transformations of the organic carbon material during the aeons that passed after the planet was formed out of dust and fire until the time when we again see organisms in the Cambrian seas. It is a reconstruction with many weak features, but undoubtedly with some positive points.

CHAPTER V
FAR AWAY

A reconstruction of the past must necessarily be based on certain observations that are sufficiently reliable to serve as a starting point. This is a trite assertion, made here solely for the purpose of provoking the question: How many well-documented data are available for a clear impression of the structure of life ever since the formation of our earth? How much is based on direct evidence, how much on indirect evidence, conditions, indications, and pure conjecture?

Our reason for raising this question is perhaps partly that we want to be sure we are undertaking a respectable investigation and not something fanciful. We know that facts as such do not guarantee scientific solidity, that a slight misconception in combining them into a theory can give quite a bizarre result. Likewise, it is clear that some imagination is required to make guesses in difficult cases. How sensible a guess is can be seen in the long run, when the idea inevitably is gradually confronted with new facts. What we are certain about here is that we are going to deal with science and not science fiction, that our attempted reconstruction will be as down-to-earth as possible, and that the interpolation of fancy needed now and then will be critically inspected. This is just parenthetical.

We can begin by analyzing the situation in the Cambrian, 500 million years ago. The information we can get from paleontologists gives us an over-all picture of life at that time as something largely confined to the seas. So far as the land regions are concerned, there is no evidence in the form of fossils from this period that indicates any advanced plant life. About 150 million years later there did exist a flora of rather low-growing plants, the main representatives of which, *Rhynia* and *Asteroxylon*, were gradually joined by rela-

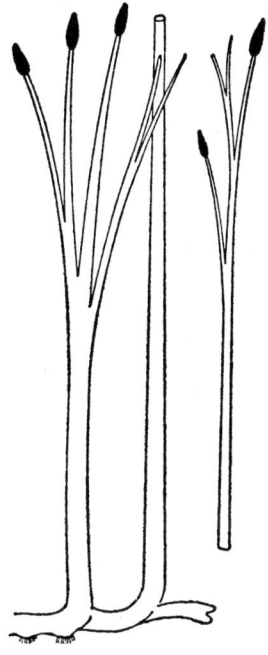

FIG. 16.—*Rhynia*

tively tall treelike structures: those that later developed into *Sigillaria* and *Lepidodendron* of the Carboniferous period, now found as fossils in coal deposits.

In itself, the absence of fossils of land plants from the Cambrian period is not proof of the complete sterility of the land regions during this period. We can ask ourselves the question: What does analysis of coal from an early period reveal about the nature of the material that was carbonized? Clearly, it was plant material, since we can often observe impressions of different plant parts in coal. On the other hand, in the fossilization process there is a constant local retention, a local deposition, of the most stable parts of an organism —what is *not* immediately metabolizable, such as protein, fat, and low-molecular carbohydrate—thus something not water-soluble or

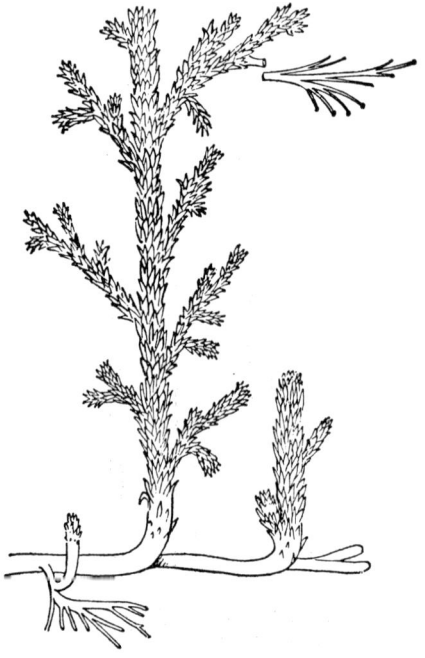

FIG. 17.—*Asteroxylon*

not easily affected by enzymes. The low nitrogen content in coal indicates, in any case, that protein (with 10 to 12 per cent nitrogen) must have been decomposed before the later stages of fossilization. What we logically conceive as the most stable part of plant material *in corpore* is woody tissue, cellulose in combination with lignin. We know from our experiences of present-day preliminary stages of fossilization—namely, the formation of peat in our peat bogs—that what we call *humus* is a complex combination of converted lignin components and cellulose, unaffected or partially decomposed. There is some indication that the formation of peat, continued for millions of years, will result in gradual carbonization, that the carbon content in the material will increase and the content of hydrogen, oxygen, nitrogen, and other elements will decrease.

Fossil material in our time is undoubtedly *the most stable type of chemical conversion product which could be formed out of the most stable part of the original material.* Wood in general, represented by cellulose and lignin in special combination, is the most stable plant material of our time, particularly in combination with resins. We can observe the structure of woody tissue in coal. There are chemical indications that coal is largely former woody tissue. Conclusion: the finds of plant fossils in the form of coal from post-Cambrian times indicate that woody plants—trees—were first developed between 400 and 300 million years ago, and that prior to this time cellulose and lignin represented a considerably smaller fraction of life than they do today.

This discussion is the introduction to an inventory of the distribution of life—as carbon—on the land and in the sea in the Cambrian period. We can establish the following: Since present-day terrestrial organisms represent 50 billion tons of carbon, of which 80 per cent, or 40 billion tons, represents woody tissue in the form of tree trunks and *the latter structures did not exist* during the Cambrian period, the total amount of terrestrial organisms calculated as carbon in Cambrian times was considerably less than it is today.

If we assume that the geographic distribution between land and sea at that time was about the same as it is today—an assumption that even with a margin of error of ± 20 per cent does not affect this argument—the upper limit of the total amount of terrestrial organisms would be 3 billion tons of carbon at the most, corresponding to what we can calculate today from the average density of the non-wooded regions on our earth. This figure, however, is obviously too high. There is no evidence of such extensive plant life during this period. The maximum figure should probably lie closer to the steppe and desert regions in our time, thus about 1 to 2 million tons of carbon. The lower limit lies at 0. At the moment we assume that the absence of preserved fossils of Cambrian land plants is synonymous with their complete non-existence, and also with the absence of corresponding fauna.

Regarding the organisms in the seas, we can begin with an entirely different starting point. As on the land, animals constituted a minor fraction of the total organic matter. The amount of existing green marine plants of all types in the sea is determined by (a) the depth to which solar radiation can penetrate and (b) available raw material for photosynthesis. Regarding point a, the conditions were probably the same in the Cambrian as they are today; in any case, the solar radiation probably could not penetrate more deeply than we can observe in our time, which is about 50 meters. It is in this surface layer of the oceans that photosynthesis takes place in all the organisms we sum up under the concept of phytoplankton. As for the raw material of green plants in general, it is carbon dioxide, water, nitrogen compounds, phosphate, sulphate, and numerous salts of different kinds. Here we come to point b: What is it that actually limits the extent of photosynthesis? It is an intricate question. It cannot be available carbonates and carbon dioxide—they as well as water exist in more than sufficient quantity: 50,000 billion tons compared to 3 to 6 billion tons of organic carbon, under present conditions. The amount of solar radiation? Hardly. The amount that during the year strikes and is absorbed by our present seas, if fully utilized, should be able to form out of carbon dioxide and carbonates up to 10^{18} grams of organic carbon = 10^{12} tons = 1,000 billion tons, which is 250 times more than we can observe now in our seas.

In recent years oceanographers have begun to understand what the limiting factor for the population in our seas is. They have localized this factor to the supply of nitrogen and phosphorus, in the form of ammonium salts, nitrates, and phosphates. Since the top 50-meter layer of sea water is, on the average, only 1/80 the total sea mass, and the nitrogen and phosphorus content here is rapidly reduced during photosynthesis, the continuing process is wholly dependent on the vertical currents which bring the nitrogen- and phosphate-rich deep water to the surface. These currents are in turn dependent on the rotation of the earth, wind conditions, and the geographic formation of the continental shelves. A complete mixing of deep water and surface water is precluded within such a short period as one year. Consequently there is a permanent deficiency of nitrogen and phosphate in the surface layers, and thereby a constant limitation of photosynthesis and organic carbon in the seas.

The conditions during the Cambrian cannot have been so different from those in our times. Weather and wind, so far as we know, cannot have changed drastically. Moreover, the nitrogen and phosphate content of the seas was probably, on the average, *lower* than it is now. Every year 100,000 tons of nitrogen and the same amount of phosphate are worn away from the land regions by erosion and carried to the seas. Assuming that this process has gone on at the same rate during the 500 million years that have passed since the Cambrian, we obtain the impressive figure of 10^{14} tons, which is practically the *present* total supply in the seas.

The last calculation must be taken with a certain reservation, but one thing is clear: the nitrogen and phosphate content of the seas in the Cambrian period can have been at the most 50 per cent of the present content, and presumably it was less than this figure. Conclusion: the amount of marine organisms in the Cambrian seas cannot have been greater than today and probably was lower. Calculated on a phosphate basis, we can estimate the total amount as about one-half that of today.

We have come so far now that we can make a comparative table of the quantitative relationships relative to life in the Cambrian and in our own time. As for the deposits of carbon dioxide and carbonates, the atmospheric carbon dioxide should have been at about the same level as it is now. The same is true of the carbon dioxide and carbonate carbon in the sea.

So far as photosynthesis is concerned, it is plausible that it should be proportional to the green plant life. Since we have here two alternatives—life concentrated wholly in the seas, and life both in the sea and to some extent on the land—the estimated figure for the average annual turnover of carbon will lie between 5 and 15 billion tons of carbon per year, approximately 30 per cent of the present value.

As an interlude here, we can sit down and catch our breath on the shore of a Cambrian landscape and ponder over the situation. The environment around us has been reconstructed with great acuity by paleontologists and paleobiologists millions of years later, with life in full activity in the sea, but perhaps particularly concentrated in the shallow water along the shores. In any case, it seems as if the higher organisms that later left traces as fossils have been rather poor

AMOUNT OF CARBON IN BILLIONS OF TONS

	In the Cambrian—500 Million Years Ago	In Our Time
Land organisms:		
Plants..................................	0–2	30–60
Animals................................	0–0.2	1–3
Marine organisms:		
Plants..................................	1–2	3–6
Animals................................	0.5–1	1–3
Total................................	~3	~50
Annual turnover of photochemically bound carbon.........................	~10	~30

swimmers and have specialized mainly in hunting food along the sandy bottom. Some jellyfish-like things are floating around in the shallower sections where primitive sponge animals have attached themselves to suitable locations. Trilobites and some predecessors of later crustaceans dominate the scene, slowly crawling around in the shallower regions, a picturesque swarm of automatism and mild excitement.

So far as plant life is concerned, we can surely assume that the shallow slopes of the shore regions are covered down to a depth of 50 meters with different types of adhesive algae—blue, green, red, and brown—as a complement to the freely floating flora of small algae. Photosynthesis flourishes and everything is teeming, eating, and thriving in a pleasant, unconscious manner. Along the sandy shores, swept up by earlier storms, lies an impressive collection of shells that gradually will be embedded, acted upon by the different geochemical conversion processes, and, in the course of time, as fossils in calcareous rock be studied by learned men.

If we glance inland, we get an impression of a rather monotonous landscape with few contours looming in the distance. The absence of trees and bushes gives the impression of a desert, and we think of the later epoch when advanced land plants will enliven the pronounced sterility. The most plausible hypothesis we can hastily conceive is that it all started as a slow invasion by the littoral algal flora; that certain types capable of surviving above the surface of the water evolved. At first they anchored under water, later in the

FIG. 18.—Calcareous algae from the middle Cambrian

moist, sandy shore, and so on, farther and farther inland, existing a short time, withering, decomposing, forming a bed of material for others to sprout, germinate, develop, and disintegrate. Following the trail of the invasion went specialized coastal inhabitants which gradually adjusted themselves to an increasingly dry environment: the non-vertebrate fauna of the land, which gradually came to compete with the vertebrates, began when a tranquil mud crawler of the lungfish type definitely remained on the land as the progenitor of primitive reptiles.

One question which we pondered over earlier reappears here: Is the inland region completely sterile? Is the land, despite the absence of stable land plants rooted in the ground through an armature of woody tissue, of cellulose and lignin, nevertheless a bearer of life in some form? Are there, at any rate, some microflora and microfauna, which co-operatively function as local life here and there, regardless of whether this life leaves behind any noticeable traces? Or, as an alternative, is the ground chemically prepared, in spite of everything, for an invasion of plant life from the coastal regions? Has previous chemical conversion of compounds left behind something to serve as substrate for later generations? After some consideration of this problem, we find that we are not sufficiently prepared to take up a position, and instead decide to discuss it again later. In the

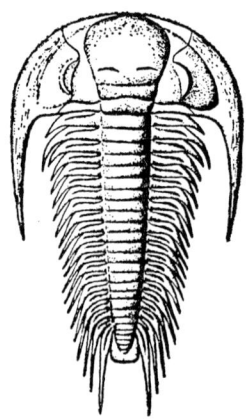

FIG. 19.—Trilobite: *Paradoxides paradoxissmus* (Wahlenberg), characteristic fossil in Sweden from the middle Cambrian.

meantime we can examine more closely the trilobite we have fished up on the shore and study its structural elements.

We find that the trilobite has certain features suggestive of the wood lice of a much later date. We find that the carapace is composed mainly of chitin, that the soft parts—after a rapid analysis—are cellular structures containing carbohydrates, fats, and proteins in the proportions that much later are standard for living organisms. We recall the fact that latter-day chemists have actually ascertained the presence of chitin and of certain protein components —amino acids—in fossil shells of trilobites and closely related animals from the Cambrian, Silurian, and Ordovician; and, further, that certain types of complex organic catalysts—porphyrins—have been recovered after having rested several hundred million years as fossil carbon material! For the formation of chitin and proteins of different types, structural elements, and enzymes, extremely intricate cellular organization is required by which carbohydrate components, fat components, amino acids, and energy-carriers of the phosphate type react with one another according to certain definite rules. The finding of protein components, chitin, and porphyrins in a trilobite or some other representative of Cambrian life is circumstantial evidence that *organization existed* for the formation of these substances in the cells of that time. The more we think about this, the more we find that there are not many alternatives of organization to choose from, *except that which is represented largely by cellular metabolism 500 million years later.* As a representative of a component of life, our trilobite seems from a chemical point of view surprisingly modern. Behind its archaic appearance is hidden a biochemical organization of cellular metabolism, the main features of which were preserved in organisms much, much later.

We are gradually beginning to see that from a biochemical point of view not so very much has happened in the last 500 million years, since the Cambrian period. We have around us in our Cambrian landscape an advanced flora and fauna, which, to be sure, has been estimated to represent only a small percentage of the number of species found in later epochs, but the already abundant variety indicates a prehistory which must have extended far back into the pre-Cambrian. How far?

We are in a somewhat puzzling situation, surrounded by this active Cambrian life. We know that Cambrian life has a total volume (calculated as tons of carbon) of about 3 to 6 per cent of the life 500 million years later; that the regions where the vital processes function most actively are the shores of the great seas—a limited area in comparison to the situation in later times. The abundance of species and also the pattern of variation are likewise considerably smaller, but at the same time the chemical components must, on the whole, be formed as something that will be maintained with stubborn conservatism for unlimited time. We find that our excursion to the Cambrian epoch has given us only a temporary observation post, and that what we are seeking lies further back in time. Some place where the extent and variability of life converge more and more toward a point? Hardly. It is a tempting thought to settle down with, but something much more difficult awaits us. As we venture further back into pre-Cambrian grounds, we are confronted with life in the form of organisms and organizations of carbon compounds that have not left behind any fossil traces. We have no morphologic details—nothing that gives the impression of organism in its modern connotation—but possibly certain remains of carbon compounds transformed into unrecognizability by the forces of the surrounding world, and possibly other material which once formed the environment for life.

It is not much, but we remind ourselves that the very concept of matter contains history. The geologic strata have much to tell us. The distribution of practically identical components of certain elements—isotopic distribution—gives some information, which, if we are sufficiently perceptive, can enable us to catch a glimpse of life-as-a-whole long before it flourished in all its abundant variation in the Cambrian.

We stretch our legs and cast a last glance around the landscape where the evening breeze is beginning to disturb the quiet peace in the environment of the coastal population. It is tempting to stay longer and become acquainted with the inhabitants and get to know their special characteristics, but we remember that we are on a journey into the past and that there is advanced literature on Cambrian life, to be studied according to our time and inclination. Our first object is to go back to the time before the Cambrian, where we probably shall encounter life in some form, but hardly anything whose external forms of expression give us the feeling of familiarity that we experienced in the Cambrian period. We leave the trilobite and calcareous algae to take care of themselves in their innocent existence in a tepid sea and hope for a reunion after an excursion into the archaic world, where life began as pure chemistry before assuming the forms which we observed 500 million years ago and which we see in modern editions around us today.

CHAPTER VI
BEYOND THE HORIZON

Behind the material world's façade of color and form, of energy and structure, we encounter the uniqueness of the elements in all their abstract complexity. It is a world of atomic dimensions, where the pervading theme is the constant interaction between positive and negative, symbolized by the nucleus of the atom with its conglomeration of positive protons and neutral neutrons, all surrounded by a cloud of negative electrons: subtle structures of organized agitation.

What gives each element in this interaction its external properties in general is the number of protons in the nucleus and the corresponding number of counteracting electrons in the cloud. This numerical distribution between the nuclear charge and the total charge of the electrons is reflected in the external properties of the atom: its degree of chemical aggressiveness, its *type* as an element per se and in combination with other atoms. The role of the neutrons in the whole atomic structure—apparently inactive units in the atomic nucleus, but with approximately the same mass as the protons although without their charge—seems to involve *stability* of different arrangements of atomic structures. If there are too few or too many in the nucleus, it becomes unstable, something that we observe as radioactive decay.

This abrupt introduction to atomic structure must be made here so that we can get an idea of what we colloquially designate *ele-*

ments. The external properties of the elements that we observe as carbon, sulphur, iron, silver, and 99 other types are wholly a function of the nuclear charge: from charge 1 representing hydrogen, to 6 for carbon, to 7, 8, and so on, to 92 for uranium, and to 103 for the latest synthetically produced new arrival in our world. The number of neutrons in the nucleus thus does not influence the *type* of the atom as such but makes a characteristic contribution to the relative weight of the atomic nucleus and, as mentioned, in suitable distribution with the protons bestows complete stability upon the atomic structure.

The supply of elements comprising the world around us, in the bottom rock of our planet, in the sea, and in the atmosphere, represents mainly stable atomic configurations, with the neutron number varying within certain limits, different for each element. Some maintain only a certain definite number of neutrons in relation to the characteristic number of protons: the relationship protons-to-neutrons has only *one* fixed value. Other elements represent an intimate mixture of several stable configurations, all of a *type* determined by the proton number but with varying *mass* dependent upon the number of neutrons in the nucleus. Gold is an example of one that we might designate a pure element; it has a certain fixed combination of neutrons and protons in the nucleus (118 versus 79, thus a total *atomic weight* of 197). In the case of carbon, we find two stable configurations of neutrons and protons in the nucleus (6 versus 6 and 7 versus 6), representing the atomic weights of 12 and 13. The atoms that represent the element gold in its stable form are thus of one and the same kind, symbolized by the formula $^{197}_{79}Au$, which here gives the *type* of the substance by the proton number 79 and its *atomic weight* 197 (the neutron number is thus $197 - 79 = 118$). For carbon we can similarly express the two constituent atomic types: $^{12}_{6}C$ and $^{13}_{6}C$, both with the characteristic proton number, which gives the *type* of carbon, but with different atomic weights, 12 and 13. The carbon we observe in different forms in our world is thus a mixture of two typical atoms with the atomic weights 12 and 13, *the carbon isotopes C^{12} and C^{13}*.

The concept of *isotope* dates from 1918, when the English investigator Aston was able to demonstrate experimentally that the hitherto known elements in many cases represented a mixture of two or more components, all of the same type for each case but with different nuclear weights, nuclear masses. With the help of the mass-spectrometer, further development of Aston's technique of characterizing an element by the distribution of its constituent atomic types with different nuclear masses—in other words, its *isotopic distribution*—has stimulated modern natural science research remarkably, not least the biological branches.

We know at present twenty naturally occurring elements of planetary matter with atomic nuclei that exhibit a tendency toward instability; we commonly call these radioactive elements. All constituent isotopes of some of these elements—e.g., uranium, $^{235}_{92}U$ and $^{238}_{92}U$—are radioactive; and the statistical probability for the decay of the atomic nucleus is, for instance, greater for $^{235}_{92}U$ than for $^{238}_{92}U$. This tendency to disintegration is expressed numerically by stating the time required for a certain amount of radioactive material—e.g., $^{235}_{92}U$—to be reduced by nuclear decay to one-half its original weight; in other words, the time that passes before one-half the constituent atoms of $^{235}_{92}U$ have disintegrated into something else. This constant, the *half-life*, is relatively long for $^{235}_{92}U$—710 million years—but short compared to the corresponding half-life of $^{238}_{92}U$—4.5 billion years.

For other elements, one or more constituent isotopes can be stable and one or more radioactive. At the end of the 1930's, the somewhat surprising discovery was made that the generally occurring element potassium, with isotopes $^{39}_{19}K$, $^{40}_{19}K$, and $^{41}_{19}K$ in a certain definite proportion, contained an unstable component, namely $^{40}_{19}K$, with a half-life the impressive length of 15 billion years. Several other elements, hitherto classified as stable, were later shown to contain an unstable isotope with a certain characteristic half-life for each case.

Each radioactive isotope shows highly individual peculiarities in the process of nuclear decay through nuclear instability. In certain cases the process leads immediately to the formation of a completely stable isotope of an adjacent element: e.g., the isotope of polonium, $^{210}_{84}Po$, by emitting particles with 2 neutrons + 2 protons, thus $^{4}_{2}He$, an α-particle, stabilizes as an isotope of lead, $^{206}_{82}Pb$. An emitted electron is called a β-particle, and here we speak of β-decay. In many cases the radioactive disintegration of an isotope can lead to the formation of another unstable isotope, which in its turn can yield another, and so on in a chain of decay until a stable end-product is formed.

In some other cases the decay can entail a branching of the process so that certain atoms disintegrate into one closely related isotope, others into another. The case of radioactive potassium, $^{40}_{19}K$, is a typical example of such a branched process: one type of decay leads to the neighboring element calcium, another to argon—$^{40}_{19}K \rightarrow {}^{40}_{20}Ca$ and $^{40}_{19}K \rightarrow {}^{40}_{18}A$—in other words, a redistribution of the characteristic nuclear charge 19 for potassium to 20 for calcium and 18 for argon.

The isotopic distribution of each element, regardless of whether the isotopes are stable or unstable, and the characteristic decay pattern of the unstable isotopes provide us with something more than a chemical-physical classification of the components of matter here on our planet and in different parts of the universe. They also give us information concerning the history of matter, and if we are sufficiently attuned to the messages about the past that lie hidden in the elements we have around us in our world of today, we have some chance of complementing our analysis of life and environment in the archaic period prior to the last 500 million years. The task is not an easy one, but there is undoubtedly a certain interest in seeing how far we can pursue this course.

The fact that radioactive disintegration takes place as a type-specific process, unaffected by all kinds of environmental influences —heat, pressure, chemical action, etc.—makes it possible to work out dating methods for longer and shorter periods. One well-known method, which has been in use since the 1930's, is based on the fact that the decay of both isotopes of uranium proceeds via several steps to the stable end-product lead. Seen as a whole, their disintegration runs from $^{238}_{92}U$ to $^{206}_{82}Pb$ and from $^{235}_{92}U$ to $^{207}_{82}Pb$. See Figure 20.

A determination of the isotopic distribution in lead, isolated from a mineral containing uranium, yields here certain information on how much time has passed since the mineral in question was crystallized out as a chemical combination. It is known that there is a stable isotope, $^{204}_{82}\text{Pb}$, of naturally occurring lead which is not an end-product of radioactive disintegration, unlike isotopes $^{206}_{82}\text{Pb}$ and $^{207}_{82}\text{Pb}$. Determination of the relationship between the isotopes 204: 206: 207 yields a measure of how much of the lead material has been formed from uranium-238 and uranium-235. Since we know the half-lives of the uranium isotopes, we can calculate with the help of the relationship between the lead isotopes, using $^{204}_{82}\text{Pb}$ as the standard, the time from the formation of the mineral—when "lead" in

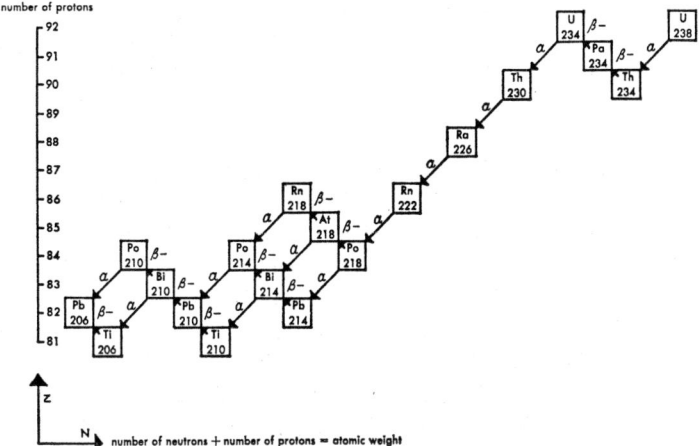

Fig. 20.—Scheme of the radioactive decay of the uranium isotope $^{238}_{92}\text{U}$. As shown, the disintegration takes place in several steps, at times via branched reactions. Each emission of an α-particle involves a decrease of 2 in the proton number. The emission of a β-particle involves an increase of 1 in the proton number and no change in the atomic weight. The end-product of this disintegration series is the stable lead isotope $^{206}_{82}\text{Pb}$. Two other similar series derive from other isotopes of uranium and from thorium, yielding the lead isotopes $^{207}_{82}\text{Pb}$ and $^{208}_{82}\text{Pb}$. Analysis of the isotopic distribution in lead, isolated from different minerals containing uranium and thorium, affords a possibility of determining the age of the mineral and—in certain cases—a chance of calculating the time that has elapsed since the formation of the earth.

the material was largely $^{204}_{82}$Pb—until now, when the analyzed lead contains a characteristic addition of $^{206}_{82}$Pb from uranium.

To sum up: With the help of the isotopic distribution of certain elements in minerals containing radioactive elements, we can obtain some information on the age of these minerals and thereby the age of the rock formations in which they are found. Furthermore, on the basis of similar measurements, we can indirectly arrive at a rough estimation of the age of the earth as a whole. This is, so to speak, the frame in which we shall fit our data concerning life and the environment for life in times long past.

The dating of material representing living organisms from past epochs with the help of radioactive disintegration in the material is, of course, wishful thinking. During the past twenty years' research, we have, however, been able to do something in this direction, even if the method is not applicable to periods further back in time than 25,000 years—an exceedingly small fraction of the past. The story is as follows.

In 1947 the American investigator W. F. Libby discovered that carbon as methane, CH_4, isolated from plant material had a certain weak radioactivity compared with the corresponding hydrocarbon isolated from petroleum as natural gas. Repeated experiments showed that the younger a carbon material was, the greater its radioactivity, something absolutely new. The effects are not great; one can reckon that 1 gram of present-day carbon has 15 atomic disintegrations per minute. What is it that disintegrates—$^{12}_{6}C$, $^{13}_{6}C$? Neither one; both are completely stable. After some consideration, Libby and his co-workers hit upon the idea that it could be the carbon isotope $^{14}_{6}C$ which, to be sure, had never been found earlier in nature but can be formed artificially by neutron bombardment of the nitrogen isotope $^{14}_{7}N$ through the following exchange:

$$^{14}_{7}N + n^0 \longrightarrow ^{14}_{6}C + p^+ .$$

One neutron in, one proton out, the mass unchanged, the charge one step lower, from 7 to 6, thus from nitrogen to carbon. Is it possible that this process, hitherto carried out only in an atomic pile, can also take place in nature?

In the upper layers of the atmosphere a slight but characteristic amount of $^{14}_{6}C$ is constantly being formed from nitrogen $^{14}_{7}N$ by neutron bombardment, in its turn a by-product of the cosmic radiation to which our earth is always exposed. This radioactive carbon isotope $^{14}_{6}C$ is rapidly mixed into the lower layers of the atmosphere, occurring here as carbon dioxide, $C^{14}O_2$, chemically so similar to the natural mixture of $C^{13}O_2$ and $C^{12}O_2$ that it has the same chance of being incorporated into green plants and of beginning the great migration via animalia, sea, and atmosphere back again to green plants. In other words, all living material receives a constant supply of radioactive carbon in the form of $^{14}_{6}C$. A determination of the half-life of $^{14}_{6}C$ has yielded the value 5,860 years. Thus, in this time the radioactivity in our organic carbon material formed 5,860 years ago has been halved, compared with a carbon material newly formed today from carbon dioxide and water. Material formed from $CO_2 + H_2O$ as long as $2 \times 5,860 = 11,720$ years ago must have one-fourth the activity of "new" material; material from $4 \times 5,860 = 23,440$ years ago contains one-sixteenth. . . . This is practically as far back as we can go, since the radioactivity is so weak in present-day material—15 disintegrations per minute per gram of carbon, dependent in turn on the exceedingly slight production of $^{14}_{6}C$ in the atmosphere—that the uncertainty in determining radioactivity in organic carbon material older than 25,000 years becomes entirely too great for the method to have practical use. On the other hand, the C^{14} technique is a new addition to archeologic dating methods, especially valuable for the period 10,000 to 1,000 years before our time.

From a general biologic point of view, it is in a way quite tragic that the only radioactive carbon isotope in nature, $^{14}_{6}C$, is as short-lived as it is. A half-life of 5,860 years is an infinitesimally small fraction of the period back to the Cambrian, not to mention the epoch prior to the last 500 million years. The guidance we have in the question of dating fossil remains lies thus in the dating of the geologic strata in which they were embedded. In itself, this is not so bad, so long as we can prove that certain carboniferous fossils were really organisms. However, if we by chance should find some carboniferous material in a very old stratum, how can we actually identify the material as "organic" if we cannot see morphologic details, i.e., something we consider typical of an organism?

So far the problem is perhaps not so much a question of dating as one of identification. Although this is a very controversial subject, we must take a position. The question is: Can the non-radioactive carbon material from epochs prior to the Cambrian leave any information on its origin—whether it was organisms or pure inorganic carbon? This question arose at the end of the 1940's and the discussion is still going on. The main features are, however, as follows.

During work with an exceedingly sensitive mass spectrometer, A. O. Nier in Chicago (1939) happened to analyze some carbon material (as carbon dioxide) of different origins. It was shown that the hitherto accepted value for the ratio between the carbon isotopes $^{12}_{6}C$ and $^{6}_{13}C$, which had more or less been considered absolutely constant, actually fluctuated somewhat, depending upon where the carbon material originated. In general, the value for carbon from calcareous rock, representing carbonates from former shells of marine organisms, was found to be around 88–90, whereas the fossil non-carbonate carbon was at the level 90–93. The same relationship proved to be valid for a comparison between carbonate-shell-carbon from present-day organisms and carbon from plant material of our time; 88–90 in the case of the former, 90–93 in the case of the latter. The situation became so interesting that several investigators began to work in this new field of research. F. Wickman in Sweden started a large series of determinations of the 12:13 relationship in carbon from plant material from all over the world, and K. Rankama, of Finland, measurements of the 12:13 ratio in carbon material from pre-Cambrian times.

The preliminary results from these investigations can be summarized as follows: a difference in the isotopic relationship between C^{12} and C^{13} *exists* in carbon material of different origins; carbon as a component of organic material has a higher ratio, 90–93, than carbon from inorganic carbonates, 88–90. Could it be that all this perhaps has something to do with the function of life-as-a-whole on our earth? Could it be a mechanism that in some way has a slight but significant preference for one of the isotopes C^{12} or C^{13}?

As a matter of purely dogmatic principle, two isotopes of an ele-

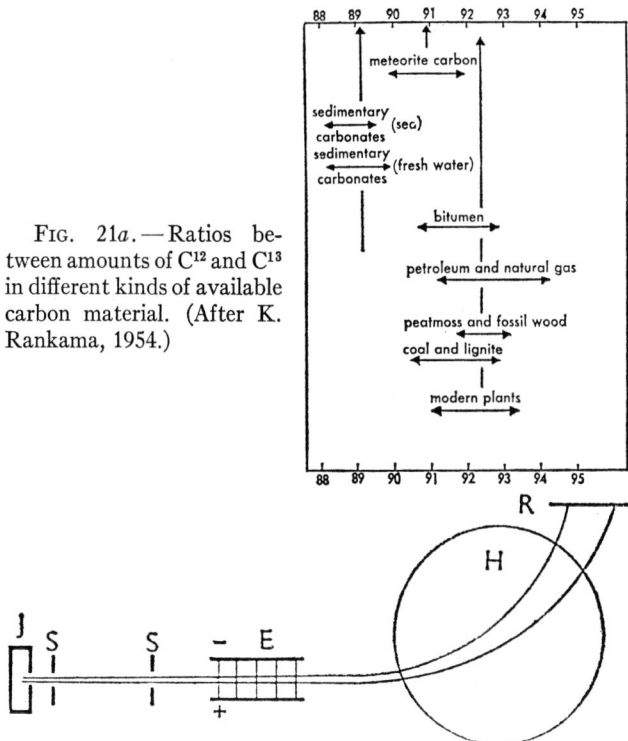

FIG. 21a.—Ratios between amounts of C^{12} and C^{13} in different kinds of available carbon material. (After K. Rankama, 1954.)

FIG. 21b.—A simple scheme of a mass-spectrometer and the separation of isotopes. From the ion-source a beam of ions passes through a combination of an electric field (E) and a magnetic field (H). By this arrangement heavier ion-particles are separated in their flight-path and collected at different localities in the instrument for registration, here symbolized by R. By this way the ratio in intensity between particles, say $12_{CO_2^+}$ and $13_{CO_2^+}$ with the total mass of 44 and 45 atomic units could be measured. In this way the isotope distribution for each element could be determined.

ment should be extremely difficult to separate with chemical methods, since the *type* is definitely determined once and for all, regardless of the *atomic weight*. As early as 1920, however, von Hevesy and his co-workers were able by means of fractional distillation to enrich mercury with some of the isotopes of higher average atomic weight than the "natural" form. The same procedure led to enrichment of potassium by its heavier isotopes—not much, but sufficient to demonstrate that a separation of isotopes was possible. H. Urey, in the United States, engaged in similar experiments to separate the two isotopes of hydrogen, ^1_1H and ^2_1H, was able to demonstrate a material enriched with the heavier isotope, a heavy water in which the component ^2_1H occurred in a concentration 300 times greater than in "natural" water. The component C^{13} was concentrated in 1940 from the natural mixture C^{12} and C^{13}, partly by physical methods worked out by F. Clusius in Germany, partly by the chemical methods of Urey and his co-workers in the United States. The same has been done with the nitrogen isotope N^{15} from the natural mixture of N^{14} and N^{15} in the relationship 334:1 up to a concentration of about 70 per cent of N^{15}.

After the concentration of certain isotopes from a "natural" mixture was accomplished, of course the question arose as to what extent living organisms might be able to carry out isotopic fractionation on a modest scale. In other words, it was desirable to determine whether there was an actual biochemical foundation for the observations that had been made concerning the $C^{12}:C^{13}$ ratio in carbon material of organic and inorganic origin.

Experiments with different photosynthetically active plants in an artificial environment of carbon dioxide in the form of $C^{12}O_2$, $C^{13}O_2$, and $C^{14}O_2$ revealed an obvious preference. Plant material—via photosynthesis and respiration—concentrates more C^{12} than C^{13} and C^{14}. The carbon dioxide given off via the respiratory processes is enriched more with C^{14} than with C^{13}, and more with C^{13} in turn than with C^{12}.

Here we can refer to the measurements of Nier, Wickman, and Rankama of the $C^{12}:C^{13}$ ratio in living and fossil plant material of the non-carbonate type compared to the same ratio for present-day and fossil carbonate. If it were a general principle that the green plant material on our earth should preferentially concentrate the lighter carbon isotope C^{12} over the heavier C^{13}, then marine organisms should expire carbon dioxide enriched more with C^{13} than with C^{12}. Consequently, the carbonates formed out of the respiratory carbon dioxide from phytoplankton and subsequently precipitated as calcium carbonate should have more C^{13} than the average carbon material of the earth in general.

A criterion for carbon material having been involved in *life* at some time—in our day or earlier—should thus be that its $C^{12}:C^{13}$ ratio deviates somewhat from the average, which seems to lie around 90–91. This latter ratio should thus actually be exhibited only by carbon material that has never passed through one or more revolutions in the carbon dioxide cycle. An ingenious way of proving this is to analyze the $C^{12}:C^{13}$ ratio in carbon material from meteorites, the material from our planetary system's stock of pebbles circling around in definite orbits, occasionally being swept up by the earth's force of attraction. Since one has good grounds for believing that the carbon of meteorite material has never been involved in any vital process, its $C^{12}:C^{13}$ ratio should be able to serve as a norm for carbon-that-has-never-been-alive. Here we have a standard:

		C^{12}	:	C^{13}	
Carbon	Meteorite	89.0	–	91.8	
	Inorganic	88.0	–	89.8	(carbonate from marine organisms)
	Organic	90.0	–	94.0	(land plants, marine organisms, and fossil non-carbonate carbon)

Thus, we should have a chance of discerning something of life in pre-Cambrian times. Carbon material from this epoch, isotopically dated as to its environment, should, with the aid of the isotopic ratio $C^{12}:C^{13}$, be able to yield information on its previous history: whether it represents carbon that was in circulation in the cycle of life in that far-distant period, or whether it was outside this cycle. In the latter case the ratio should be around 90; in the former, either in the region 91–93 or 88–90.

A series of measurements from the Rankama group in Finland

provides reason for reconsideration here. Some carbon material from the Karelo-Finnish S.S.R., as a mineral called Schungite, is dated on certain geologic grounds to an age of approximately 1,000 million years. At the same time its $C^{12}:C^{13}$ ratio is stated as 90.55–92.9. Something that is labeled as anthracite from the pre-Cambrian period, found in the Siksberg Mine at Grythyttan, Sweden, shows a ratio of 92.4. Pre-Cambrian graphite from certain localities in Sweden and Finland has yielded ratios between 90.0 and 91.7. Furthermore, something is found in Swedish granite that is called mineral pitch, an amorphous, apparently bituminous carbon material that has been dated to the pre-Cambrian. Two ratios that have been determined for it lie between 91.9 and 92.3.

If this reasoning holds good, we should have here the remains of former organisms living in an environment where photosynthesis took place with accompanying slight, but obvious, isotopic separation. Another fossil of something is a piece of carbon material found by the geologist Sederholm in Finland in 1909 in the region of Tammerfors. It is carboniferous and has a structure that is reminiscent of an organism of at least 1,200 million years ago. Rankama's measurements show a ratio of between 90.0 and 92.0. Is this also a case of the surviving carbon remains of an organism from the pre-Cambrian?

The question is open. Additional evidence comes from an entirely different quarter. During recent years Thode and his co-workers in Canada have studied the isotopic distribution in sulphur from different minerals and material, inorganic and organic, present-day and prehistoric. Thode finds a variation in the ratio between the sulphur isotopes, $S^{32}:S^{34}$, which can be traced back to sulphur metabolism in general in living organisms. The ratio in present-day sulphide material is 22.95–23.05. In prehistoric sulphide minerals from 1,500 million years ago, the ratio is 21.90–22.08. In sulphate minerals from our time, the ratio is 21.60–22.12 according to some determinations that have been carried out; and in sulphate minerals from 800 million years ago the ratio lies between 21.88 and 22.18. A conclusion can be drawn that prior to 800 million years ago there was very little biologic fractionation of the sulphur isotopes S^{32} and S^{34}. The ratio was the same for sulphide and sulphate. After this time an obvious difference is detected between the ratios in sulphide material and sulphates. This alteration of the environment can, according to Thode, hardly be anything else than the shifting of early organisms from an anaerobic mode of living to an aerobic, with the concomitant production of oxygen. In other words, we have a slight chance, but a chance nevertheless, to date the origin of modern photosynthesis from around 800 million years ago. This gives us a glimpse of something beyond the horizon, of life long before organisms became

FIG. 22.—Drawing of fossil carbon material from Finland with certain characteristics indicating that it may be carbonized organic material, as shown here by the four outer contours in the section of the stone material. Sederholm, who discovered the structures, called it *Corycium enigmaticum*, intimating a certain doubt that the material could have originated from a pre-Cambrian organism. Rankama's measurements of the distribution of the carbon isotopes C^{12} and C^{13} lend some support to the possibility that this is an exceedingly ancient organism, the outer contours of which have been preserved here. Analogous material with an age of about 1,500 million years has been found in other localities.

so substantial that their outer contours could be preserved as fossils. In accepting these data we can assume—as a working hypothesis— the existence of organisms in an anaerobic environment back about 1,500 million years, which in turn is supported by the $C^{12}:C^{13}$ measurements.

We have penetrated as far back as we can into the past on the basis of our knowledge of the chemical signs of life of organisms in our time, in the Cambrian, and, by indirect reconstruction, in the pre-Cambrian. We cannot expect anything more from the last mentioned than sparse indications of the existence of life—without any details concerning its formation—when something existed that could actively alter the isotopic relationship of the carbon material and sulphur compounds of the earth's surface. The conclusions we can draw here concerning the life and mode of living of organisms in the pre-Cambrian period are naturally limited by great uncertainties; nevertheless, further development of isotope studies offers certain possibilities of obtaining a better grasp on this complex problem. Every year brings fresh suggestions within this field of research. An interesting variant of the study of the isotopic balance between C^{12} and C^{13} in the question of carbonates in equilibrium with dissolved carbon dioxide

$$C^{13}O_2 + C^{12}O_3^{--} \longleftrightarrow C^{12}O_2 + C^{13}O_3^{--}$$

exists with regard to the equilibrium

$$CO_3^{16--} + H_2O^{18} \longleftrightarrow CO_3^{18--} + H_2O^{16}.$$

In the latter equilibrium reaction there is an exchange in the water content of the oxygen isotopes O^{16} and O^{18}. The equilibrium of both reactions is *dependent on the temperature at which the reactions take place.*

By measuring the ratio 12:13 in regard to carbon and 16:18 in regard to carbonate oxygen, one can obtain some idea of the temperature at which a certain equilibrium became frozen, so to speak. One can actually determine fairly accurately the temperature at which a certain carbonate material was once deposited as a fossil in the sea in far-distant times. Urey and others have used this method to determine the temperature of the sea in epochs many hundred million years before our time. The results indicate that back to the Cambrian there were not any great variations in the external environment of marine organisms, and that the dynamic carbon dioxide equilibrium which we observe today as a link in our modern edition of the carbon cycle was established far back into the pre-Cambrian.

At this point, however, the possibilities of describing chemical history begin to be exhausted. Now we shall attempt to analyze the consequences of the fact that once upon a time a simple carbon material was deposited around our earth in the first epochs of its existence as a planet in a newly formed solar system. We shall start from what we can reconstruct of conditions at the dawn of time and proceed *forward* instead of backward. Our goal is to link, through aeons of chemical activity, both excited and sluggish, the action within prebiotic carbon material with what we interpret as life in the pre-Cambrian, which later in the Cambrian seas exhibits striking features of flourishing vitality. The theme of our investigation of the abundant chemical possibilities of the first billions of years will be based on certain reasoning applicable to chemical reactions of today. Such a reconstruction of an archaic past must naturally involve many moments of uncertainty, but risks must be taken when one sets out on an adventurous enterprise.

CHAPTER VII
OUT OF DUST AND FIRE

Out of timeless uniformity something became differentiated out of something—something-after out of something-before, fluctuations out of stillness, matter out of pre-matter, the emergence of time and distances, amplitudes and frequencies, structures and properties—all crystallized into the uniqueness and distribution of the atoms: a network of forces that further developed into the potentialities for complication which is a property of the elements.

The basic theme in our conception of the conditions from which our planetary system was once formed is the idea of local heterogeneities formed out of the homogeneous; differentiated matter out of undifferentiated; accumulations of matter out of uniformly distributed matter. What we see around us are variations of this theme: the abundance of subtle variation in our closest surroundings, the distribution of matter in our planetary system, the morphology of the galaxies, the structure of the universe. In our primitive attempts to reconstruct the formation of our planetary system, our minimal home region within the whole, our thoughts circle around the possibility that out of a homogeneous cloud of gas and dust certain particles grew at the expense of others; that an increasingly accelerated chain reaction gave rise to the very great out of the very small, our sun out of the rest of the dust.

Whether our planetary system, as we see it today, was formed at the same time as the central body, with its enormous accumulation

OUT OF DUST AND FIRE 87

of matter, is an open question. In all probability, however, the terminal phase in the central accumulative process must have had a fundamental influence on the rest of the surrounding matter. As we now imagine the whole process, more as a vision than a theory, the central region—composed mainly of hydrogen, like the cloud as a whole—was in the beginning a local accumulation of increasingly denser gas containing a small amount of dust particles, and later contracted into dense matter with a total weight the order of 10^{28} tons. At a certain critical compression of this mass, which in the center must have had a pressure of billions of tons per cm^2, and at a temperature which must have increased at an accelerated rate during the compression, a specific confrontation between certain atomic nuclei took place in the center. A thermonuclear reaction was started with the coalescence of hydrogen isotopes into helium, liberating awe-inspiring energy and drawing the closest surroundings into the same fusion process. As a cosmic conflagration, the process must have spread through the orb out toward the surface, which after an extremely short time flared up in stellar white heat: a furious event with a subsequent radiation of light and particles that after a couple of hours reached out to the diffuse limits of the cloud at the edge of emptiness.

This combustion of the central body into our sun, with its radiant energy, particle radiation, and solar electromagnetic field, must have affected changes in the distribution of matter within a radius of at least 60 million miles. Furthermore, we know that the matter is distributed in a characteristic way, with Mercury, Venus, Earth, Mars, and the small planets and satellites representing hardly one-tenth of one per cent of the total material. The main part is localized to the regions represented primarily by Jupiter and then decreasingly by Saturn, Uranus, Neptune, and Pluto. Was this distribution already fixed when the sun became a sun, or did the sun during the first millions of years of its existence exercise an organizational influence on the surrounding matter?

The conceptions of this phenomenon are somewhat divergent. A few investigators, such as von Weizsäcker in Germany and Kuiper in the United States, see the original gas-dust cloud slowly differentiated in regions of concentrated matter in a system of organized turbulence, from which local concentration processes are formed, both a central mass—later sun—as well as diffusely limited protoplanets in distribution around the central body. All these formations are further compressed into more compact material structures, and such a high mass density of the central body must sooner or later lead to a thermonuclear chain reaction. In other words, both the sun and the planets were formed as consequences of the same differentiation process. Others, for example Alfvén in Sweden, are inclined to believe that once a sun has been formed, its mass, radiation, and magnetic field are capable of organizing surrounding undifferentiated material into relatively defined regions, the formation and composition of which can be determined by the distribution of elements in different combinations in the original material. According to the working hypothesis of Alfvén, there is a definite possibility that the sun during an early period passed through local cosmic accumulations of matter a couple of times and organized this undifferentiated something—gas molecules, more or less charged, and molecular aggregates, of microdimensions up to particle size—into concentric regions. Local concentration processes of these regions should lead to growth of larger particles, then to still larger: a gradual accumulation of the regional material into larger units, planets. By assuming a repeated confrontation of our sun with cosmic accumulations of diffuse matter of different types, one can explain why the material in our planets is as different as it is.

We must gradually take up a position regarding the question of the composition of the material that will definitely constitute our earth, thereby getting an idea of the original structure of the carbon compounds that later, in the course of time, become components of life on our planet. In what form was it accumulated in the first period of planetary existence—this reaction-product of interstellar matter slowly cooled in an excess of hydrogen? Did the earth receive a particularly developable fraction of matter compared to the neighboring planets; or was the distribution the same, and did subsequent local transformation lead to a special development?

To begin with, we can glance at the composition of the matter that now exists in the fiery systems represented by suns of different

OUT OF DUST AND FIRE   89

types, molten comet material, and interstellar material in general. We can make some comparisons here with the composition of matter from the relatively cold systems represented by the earth and its neighboring planets. See the following table.

Atoms and Molecules in Very Hot Systems		Atoms and Molecules in Relatively Cool Systems	
s */ /	H, H⁺; He	H₂; He	Jp, Sa, U, N?
s * // k	CH, CH⁺	CH₄, etc.	Jp, Sa, U, N
s * // k	C₂, C₃	CO	Jp?
s * k	CO, CO⁺	CO₂	V, E, (M)
s * // k	CN, CN⁺	HCN? (CN)₂?	Jp? Sz?
s *	NH	NH₃	Jp, Sa
s *	SiH	N₂; SiO₂	E; all planets?
s * // k	OH	H₂O	E, (M), Jp?
s *	SH	H₂S; SO₂	Jp? Sa?
s * // k	Me⁺, MeO⁺, MeH	Me, MeS, MeO "mineral"	All planets

s = our sun
* = other suns
// = interstellar material
k = hot comet material
E = Earth
V = Venus
M = Mars

Jp = Jupiter
Sa = Saturn
U = Uranus
N = Neptune
( ) = presence in small amounts established
? = presence possible but not established

As far as stellar material on the whole is concerned, including the material in our own sun, the general composition is similar to what it was in the earliest period of the existence of the universe. Naturally every sun has undergone considerable change in its composition since that time through all kinds of nuclear reactions on a large scale, but the picture we have of atoms and molecules in great thermal agitation in our sun of today is a reflection of the most probable condition for the same material during the first developmental phase of the planetary system. We have a hot environment of hydrogen as the dominant element in the forms H and H⁺ and in combination with other elements, *hydrides*. Carbon exists as ionized hydrocarbon combinations, CH⁺, CH₂⁺; or neutral radicals, CH, CH₂; and also as the carbon-nitrogen combinations CN and CN⁺. Further, we find that the combination carbon-carbon as C₂ is common and that the carbon-oxygen compounds CO and CO₂ can be detected, most frequently ionized to CO⁺ and CO₂⁺. The dominating form of carbon is CH and CN. We find this combination in spectra of comets, material that since the earliest years of the planetary system has occasionally entered and re-entered the region close to the sun, temporarily heated and thereby giving us a chance to estimate the composition of cosmic matter from the past. Someone has described comet material as a fossil from the time when the planetary system was formed. The same is true of the larger planets, where we can estimate the lighter element content of the outer contours by means of spectroscopy. Jupiter, Saturn, Uranus, and Neptune, as well as some of their satellites, all exhibit the combination CH_4, the hydrocarbon methane, as a component of the outer layer of the atmosphere. The first two contain also a certain amount of hydrogen-bound nitrogen, NH^3, ammonia. Otherwise hydrogen and helium are probably dominant—as a whole, a picture of "frozen" chemical equilibrium from a period when hydrogen was the dominating element in the material that became *both* sun and planets.

Astrophysical analysis of spectra of interstellar matter—the material that in exceedingly thin concentration fills up the space between the suns in our universe—reveals that here also hydrogen in different forms dominates and that the existing carbon occurs as the combinations CH, CH^+, CN, C_2, and C_3. This is another indication of a composition that has not altered much during the last five billion years.

In general, the atmosphere of the giant planets, Jupiter and its companions, exhibits the inevitable result of a *cooling* of originally existing carbon compounds in an atmosphere with hydrogen the dominating element. In like manner, the composition of the carbon compounds in our sun suggests the composition of the original material before it was heated up to that degree of thermal agitation that is characteristic of a sun. From both directions we can obtain arguments for a reconstruction of the nature of the material that at one time became our earth. Regardless of whether the condensation processes worked locally with the same material in the inner and outer regions of the planetary system, or whether a regional concentration of one or the other type of material might have taken place under the influence of the sun, the primary substance of the earth must have contained carbon *in the form of hydrocarbon and carbon-nitrogen combinations as dominating components*, compared to the carbon-oxygen combinations CO and CO_2.

We have arrived at the first stage in our analysis of the condition of the earth as a young planet. We can see its incipient condensation as a faint copy of what took place in the formation of the giant planets, not to mention the conditions in the formation of our sun, with the subsequent thermonuclear cataclysm representing the price that must be paid for great accumulation of matter. Let us familiarize ourselves with the situation five billion years ago, and let our time-unbound intellect investigate the chemical possibilities available on the surface of a smaller planet newly formed by the aggregation of cosmic dust.

The young earth plows its course around a newly arisen sun. Each revolution accumulates more and more of the cosmic froth, particles of hydrogen, water, hydrocarbons, and dust representing metallic oxides. The whole process is one of growth, an unorganized sweeping-up of all the debris in its path, eventually reaching a final stage where no, or at least very little, migrating material remains to be collected in the region. As the planetary mass grows, its gravitation increases. The rate at which it captures material increases, as does the rate of energy liberation in the form of heat with momentary light phenomena, white incandescence, and chemical changes. The moon follows its silent orbit, exposed to the same continuous bombardment to a lesser degree. The heat increases for both bodies, but more for the earth than for the consort it "captured" by chance at some earlier time and now is indissolubly united in eternal combination with it, the minor component of a double planet. Each accretion of new material is followed by consequences which can be set up in equations to obtain a certain understanding of the factors involved in an aggregation process, especially valid for the liberation of gases from an orb during growth,

$$V_e^2 = \frac{2GM}{R}; \qquad V_m^2 = \frac{3kT}{\mu}; \qquad T = \frac{2GM\mu}{3kR} \cdot \frac{v_m^2}{V_e^2}$$

where M = mass of a spherical body
R = its radius
T = absolute temperature of the outer gas envelope
G = gravitation constant
k = a constant = $6.67 \cdot 10^{-16}$
μ = mass of a certain gas molecule
v_m = its mean velocity within the gas envelope
V_e = escape velocity—i.e., speed required for leaving the sphere $(v_m > 0 \cdot 2V_e)$.

These formulae say that the thermal effects for each incidence on the surface increase proportionally with the square of the radius of the planetoid and that an elevated temperature reduces the chances of retaining substances of low molecular weight. On the other hand, the gravitation of the sphere increases in proportion to its mass and thus counteracts the tendency of low molecular substances to leave the surface. These two effects interact in an intricate manner in the last phase of the formation of the planet, and it is difficult to calculate the surface temperature of the planet in its final stage of growth. The difficulty lies in the fact that we have so few possibilities of estimating the velocity of the aggregation in the last phase. The more rapid the accumulation of material, the hotter the surface; the slower the accumulation, the greater the chances that some of the heat can have escaped by radiation, thus producing a cooler surface. At this point views diverge; the majority of those who have worked on the problem believe that the surface of the earth, at least during a short period, has been molten with a temperature of up to 1,500° C.; others assume a considerably lower temperature.

My own opinion is that in the final period of the growth of the earth, the increased gravitation must have led to an accelerated accumulation of material with a concomitant elevation in temperature which must have reached its maximum when the mass of the earth was about 90 per cent of the present mass. At the same time, the orbital material available for accumulation began to be exhausted. The incidences per unit of time became fewer, the thermal radiation from the planetary surface began to have a chance of catching up with the heat resulting from the incidences. The temperature fell, and when the last per cent of cosmic matter was added to the earth, the surface was relatively cool, maximum 200° C., a figure that must be taken with reservation. One thing that must not be overlooked in this connection is that the process is still going on to some extent. Certain observations suggest that the earth in our day still accumulates more than ten tons of cosmic matter per year, remains from the great accumulative process of five billion years ago.

If we attempt to familiarize ourselves with the situation on the surface of the earth when the aggregation process was largely completed, we note at first the presence of a turbulent atmosphere over a restless surface. As a general observation at this stage, we can ascertain that the atmospheric constituents are dominated by the hydrogen-bound elements; let us for the sake of simplicity classify them as *hydrides* of different kinds, a consequence of the cooling and concentration of light elements from their pre-planetary forms of existence. In chemical symbols:

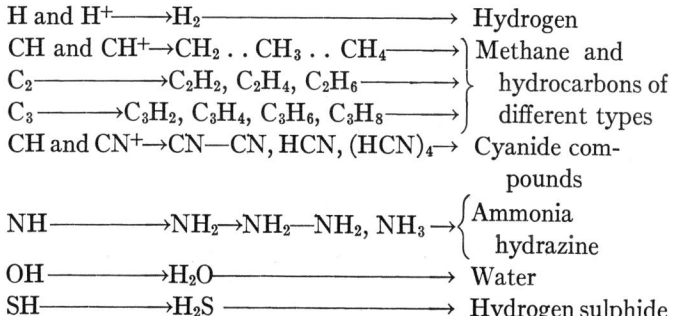

From this hydrogenated material molecular hydrogen must have escaped relatively rapidly out into space, because of its low molecular weight, 2. Other hydrides were thereby concentrated into an increasingly dense atmosphere, with an increasingly rising temperature, toward the surface, where the confrontation with the components of the surface material initiated a chemical conversion process on a large scale. We can reckon with the solid and flowing material of the earth's surface having been largely chemically stabilized at this point as a conglomeration of metals, metallic oxides, and metallic salts, with oxygen in the bound form as the dominating feature. In the first accumulative phase, this material, collected in a sort of thin mixture, must have also acquired several hydrides of different kinds. As the orb was compressed, the mixture became increasingly warmer, and a general reaction between hydrides and oxides must have been inevitable, a reaction that we can symbolize as follows:

$$2MeO + RH_2 \longrightarrow 2Me + H_2O + RO.$$

The result: water and other oxides under extremely high pressure and high temperature, which must occasionally have had an outlet of cataclysmic dimensions and led to an incipient concentration of gaseous oxides in the atmosphere. In addition to this chemical incitement to general unrest in the inner and outer regions of the sphere, compressive processes of different kinds must have led to great settlements. Earthquakes and volcanic eruptions must have been the rule rather than the exception, with molten material exposed now and then in dramatic intensity.

All these local regions of incandescence and heat became for a long time thereafter the meeting place for the hydrides of the atmosphere and the minerals of the surface layer, where the general reaction between metallic oxides and hydrides must have dominated. Seen as a whole, the process looks like this:

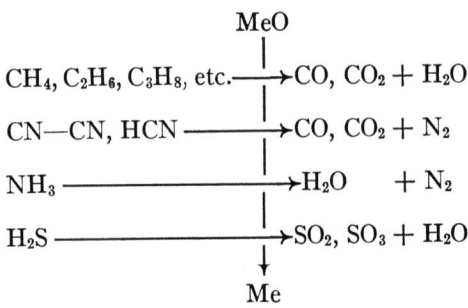

In other words, the early confrontation between hydrogenated material of the atmosphere and oxides of the surface layer at a temperature of over 200°–1,000° C. must have led to a slow reduction of the atmospheric *hydrides* and a slow increase of the gaseous *oxides:* carbon monoxide, carbon dioxide, water, sulphur oxides, and nitrogen as N_2. As the temperature in the different border regions between atmosphere and bedrock increased, a transformation of another type than this general oxidative reaction also must have occurred, namely, thermal dehydrogenation and cleavage of more complicated hydrocarbons into simpler ones:

$$\left.\begin{array}{l} C_2H_6 \ldots \ldots C_2H_4 + H_2 \longrightarrow \\ C_2H_4 \ldots \ldots C_2H_2 + H_2 \longrightarrow \\ C_xH_y \ldots \ldots C_xH_{y-a} + \tfrac{a}{2}H_2 \longrightarrow \\ C_xH_y \ldots \ldots C_zH_n + C_sH_m + H_2 \longrightarrow \end{array}\right\} \begin{array}{l} \text{Leakage from} \\ \text{the planet} \end{array}$$

The above symbolizes the whole in an exceedingly simplified form. The content of all this, however, is a process that has certain features in common with modern high-temperature treatment of hydrocarbons—so-called cracking—whereby complicated hydrocarbon constituents from petroleum are converted into simpler ones suitable for fuel. The hydrogen that is liberated as a by-product in this type of process can either escape into the atmosphere and later leak out from it or be oxidized into water on the way.

If we should attempt a summation at this stage, we might say that the boundary surface between atmosphere and bedrock in the earliest developmental period of the earth was the center for *a tendency toward oxidation and simplification.*

What happens to the oxidized and simplified material which at this stage of chemical violence continuously escapes to the atmosphere? So far as the oxides are concerned, water as well as carbon dioxide, plus nitrogen, at the proper distance from the hot surface will be relatively stable, sluggishly reactive. Dehydrogenated carbon compounds, on the other hand, have a pronounced tendency to be reactive, to form combinations with different components in the environment. It is highly probable that in the cooler regions of the atmosphere this tendency appeared at an early stage, leading to a rich flora of more or less complicated molecular types. Furthermore, ultraviolet radiation must have played a significant role as activator for sluggishly reactive material by its ability to form reactive radicals of different kinds, leading in turn to new combinations. Electrical discharges in the atmosphere—and this period certainly produced monumental thunderstorms—have the same effect: electron-removal, forming ions and radicals. In fact, some interesting model experiments have been performed: a gaseous mixture assumed to simulate the early atmosphere was subjected to electrical discharges and ultraviolet radiation. In the first reported experiment, carried out by S. L. Miller in the United States in 1953, a mixture of methane, ammonia, water, and hydrogen was subjected to discharges for twenty-four hours. It was later possible to isolate several fairly advanced new combinations from this material, including a number of amino acids of the general type that occur as building blocks in proteins of today—a complete surprise!

One thing worth recalling in this connection is the presence of cyanide compounds in the earliest atmosphere, in themselves an incitement to complications. Both cyanogen, CN-CN, and hydrogen cyanide, HCN, have a natural inclination to autocondensation during the formation of higher molecular species. A tetramerous form of hydrogen cyanide, $(HCN)_4$, with the technical name 2-amino-1-iminosuccinodinitrile, undergoes in the presence of water at high temperature a cleavage into glycine (an amino acid) and oxalic acid, both products that are found in nature today. Furthermore, hydrogen cyanide can be combined with so-called aldehydes and ammonia into compounds which later can be split by water into amino acids. There is a definite possibility that the formation of amino acids in Miller's model experiment goes back to an intermediary formation of hydrogen cyanide, with attendant complications in all directions; such a tendency is pronounced in aqueous solutions but also can take place in a gaseous phase under suitable conditions.

Another reaction that we can consider is the formation of elementary sulphur and sulphuric acid in the early atmosphere. The former process is a consequence of the reaction between primarily existing hydrogen sulphide and secondarily formed sulphur dioxide, the latter a product of water and sulphur trioxide. Hydrogen sulphide has, further, a tendency to react with cyanide compounds, resulting again in a great number of complications. An exact summation of all conceivable possibilities in the confrontation between the components of the original condensed atmosphere and the substances formed in the hot regions of the surface that returned to the atmosphere would easily fill a technical monograph of advanced organic chemistry. We can, however, summarize our impressions of what took place in early times in the higher layers of the atmosphere by expressing the general tendency as something leading to *deoxidation and complication.*

Turbulence in the atmosphere forces some material into renewed contact with the destructive forces of the earth's surface. Again we see oxidation and simplification, later compensated by deoxidation and complication in the cooler regions of the atmosphere. Carbon material migrates between two poles of counteracting tendencies. The first accumulations of material are slowly consumed during the formation of something else. Simplicity becomes multiplicity in the continuous cyclic process of destruction and reformation. Something is taking place, accelerated by the differences in energy between hot and cooler regions and activated by the ultraviolet radiation of the sunlight—turbulent reactions of an enormous number of carbon compounds in continuous transformation, vaguely suggestive of something we shall encounter a couple of billion years later. Life? Yes.

It is this flux of carbon compounds that will continue indefinitely from now on. Already developed in principle as a simple chemical theme, its variations will, during the course of time, become increasingly advanced. The accelerating force will gradually assume a different character; the dualism between the hot and the cold will be replaced by the interplay between chemically active and chemically sluggish regions and then by the interaction between light-activated hydrogenation and destructive dehydrogenation. Atmosphere and surface will gradually undergo alterations of structure and composition; the participants in the interplay will become increasingly complicated, specialized. Out of fire and dust, cosmic material aggregated into a small planet, and the innate reactivity of the material —a legacy from the beginning of time, space, force, and structure— was permitted to flourish uninhibitedly into abundant variation.

The first glimpse of life in all its chemical primitiveness that we have seen here could be described as the prologue in a classical drama, introducing the plot and the actors. The introduction suggests the whole, its details merging into an impression of something evident, something inevitable. As the play unfolds, the theme is developed in all its epic force. We have obtained a vision of the plan of the whole, of the actors in the first act of the play. We have a vague feeling of a continuous exchange of roles and that we ourselves sooner or later shall be called on the stage for a few lines. This, however, is a long way off. Let us see how the whole thing develops. We have plenty of time.

CHAPTER VIII
PATIENCE

Gradually the surface begins to cool off. Eruptions of red-hot lava become more and more infrequent; seismic unrest begins to be replaced here and there by crystallized silence. It is still hot, more than 200° C., and the components of the atmosphere are kept in constant motion, continuous thermal winds, incessant radiation of heat from the upper layers, resulting in intermixture, chemical reactions, discharges. A certain regional thermal distribution begins to prevail, changing from one million years to the next, becoming more accentuated as the temperature falls below the 200° C. mark. The sun, which earlier was shrouded by the haze of a smoke-filled atmosphere, now begins to be obscured by cloud formations. Less volatile atmospheric constituents condense. It begins to get cloudy. Drops of water reach the ground, only to be immediately evaporated and returned to the atmosphere; but dissolved components of high molecular or aggressive nature remain below. The water vapor condenses and falls again as drops of moisture: a washing-out effect. Carbon compounds of different kinds follow along with the whole process, up and down, sometimes accumulated on the ground, sometimes destructively converted in the hotter regions, sometimes only making a round-trip. It is overcast. The first rain falls—to remain on the earth.

How much water was there in the atmosphere at this time, a few million years after the aggregation of the earth as a planet; i.e., before water existed as a liquid on the surface of the earth? If we assume that all the water now present, 1.8×10^{18} tons, was uniformly distributed as a gas around our earth (with a surface of 510×10^{16} cm^2), the pressure per cm^2 would correspond to approximately 300 kg, which could exist at a temperature of 350° C. We know, on the other hand, that water is still being given off to the atmosphere even today and, further, that water must have been formed continually in the constant oxidation of pre-planetary, hydrogenated material in the earth's first epoch as an individual planet. It is impossible to give any exact figure; but by way of estimate, the total amount of water a few million years after the formation of the earth should have been about one-tenth the present amount, which gives us a pressure of 30 atmospheres, at a temperature of 180° C., before we can begin to reckon with rain on a large scale.

Then came the deluge. A rain of incredible intensity slowly filled the valley cavities in the early surface. A rain of dissolved constituents, a purification of the atmosphere, a formation of lakes, of seas with acid water, in an increasingly clarified atmosphere of nitrogen, nitrogen monoxide, carbon monoxide, water vapor, hydrogen sulphide, and inert gases as the dominating components. A couple of million years later the sunlight began to break through the clouds, revealing a clear sky of greenish blue, unlike our present one, which is of a far later date. The landscape was enchantingly illuminated by sunlight with a pronounced tinge of ultraviolet, creating intense and varying colors, at a distance grayish blue in the moisture-saturated atmosphere, occasionally with isolated signals of fantastic fluorescence.

How much of the landscape was above the surface of the water at this point is an open question. Probably such differences in level that we can observe today between the highest mountain peaks and the greatest sea depths did not exist at the time when the sea was a novelty for the planet. The average depth of the sea in our time is 3,800 meters; the average altitude of the land, 825 meters. Even if we assume a water mass at that time corresponding to an average depth around the earth of 380 meters, one-tenth the present, it is hardly probable that any greater land masses during the first period were exposed above the surface of the sea. We have to consider that the first rains were of such furious intensity that during the one or more millions of years of their duration they eroded the young landscape with extraordinary force. This took place at such a rate that the folding of the mountain ranges—if such a process actually occurred at this early period—could hardly counteract the leveling tendency of the acid precipitation, with its destructive dissolving effect.

Our vision of a fresh earth, awakening after a forcible cleansing of the atmosphere and bedrock, exposed to sunlight and accentuated alternation between night and day, presents a picture mainly of seas with larger or smaller solitary islands breaking the monotony of the horizon.

With the formation of the first seas, the very character of the flux of carbon compounds must have exhibited new aspects. Through the washing-out of the atmosphere the early sea must have been a weak solution of the more complicated material; left in the atmosphere were primarily nitrogen, besides some simple hydrocarbons with methane as the dominating component, some carbon monoxide and carbon dioxide, plus a small fraction of inert gases, very much less than today. There was in addition naturally a great amount of water vapor, which stands in a certain relationship to the temperature of the seas.

How much material was dissolved in the seas in this period? We should be very cautious here in our estimation. To begin with we can go back to the conditions in the Cambrian, when, as we saw earlier, the amount of total carbon as living organisms must have been about 1/25 the present amount—i.e., about 2 billion tons. The mass of the Cambrian seas must have been of the same order of magnitude as the seas of today: about 10^{18} tons. Together these represent a concentration of less than 0.000001 per cent, one millionth of one per cent, a minute dilution. Even if we assume that under the conditions existing when the sea was young the total mass of water was less than today, perhaps only one-tenth the present mass, we arrive at a concentration of carbon compounds around 0.00001 per cent. We have hardly any right to assume a stronger concentration; in any case, it is difficult to find any tenable argument for increasing the figure 0.00001 per cent to, let us say, 0.001 per cent. If we should amuse ourselves by assuming a concentration of 1 per cent of non-carbonate carbon, this would correspond in round figures to 10^{16} tons, an absurd value that represents the sum of *all* carbon—carbonates and non-carbonates—in the uppermost 10 kilometers of the crust of the whole earth.

Thus we must accept the fact that, from whatever point of departure we examine the estimated concentration of non-carbonate carbon—i.e., organic compounds in general—in the first sea, it must have been below 0.00001 per cent. This weak solution represents at this point all complicated carbon material that in earlier epochs was formed in the interaction between atmosphere and bedrock. The simplified material now occurs distributed between the atmosphere and the ground, with carbonic acid and carbon oxides both in the atmosphere and dissolved in the seas, and nitrogen in the atmosphere as well as some hydrocarbons. The interaction between breakdown and synthesis, which in the foregoing epoch must have represented a chemical furioso, has now slowed down to an adagio.

Many investigators—especially Oparin in Russia, who has devoted great attention to the question of the chemical processes in the newly formed sea and who is above all an authority on the subject of life in its earliest formation—frankly assume a considerably higher concentration of dissolved organic compounds than I myself venture to accept as a working hypothesis. According to Oparin, a great many chemical processes should occur in the new sea, leading to increasingly complicated substances, processes which in themselves are very plausible but which appear to me to have very little chance of taking place on a large scale in a dilution greater than 0.001 per cent. So far as I can judge, Oparin's ideas are probably much more valid if we assume an earlier concentration of the material. If such was the case, where and how?

It seems probable to me that in the period under discussion, concentration could have taken place along the shores of the seas, where wind and waves occasionally must have washed up a considerable amount of the weak solution that was later dried out by the heat of the sun. Moreover, we know that the wind today carries a slight but nevertheless significant amount of salt, blown off, so to speak, the wave crests and borne inland. Rachel Carson states in her book *The Sea around Us* that a certain region in India about 100 kilometers inland receives yearly 60 tons of sea salts per square kilometer. If we familiarize ourselves with processes of this type that have been going on for a couple of million years, we can with some justification assume slowly growing organic carbon compounds along the coastal regions of the early sea. We have other grounds for this assumption from investigations that have been recently carried out to determine the amount of nitrogen compounds that have been transported inland from the sea by wind and weather. Slowly the fringes of land must have been thus enriched with organic material from the sea. It cannot have been very much, but, nevertheless, enough that the rain over the land regions, which later became brooks, streams, rivers, must have contained a concentration of simple carbon material—lower, to be sure, than the sea itself but sufficient for subsequent concentration in shallow lagoons and inland seas.

Now and then inland seas must have been slowly cut off from outlet to the seas, somewhat similarly to what we can observe today throughout the world. The Caspian Sea and the Dead Sea are only two examples. *Here* there would have been opportunities for a concentration process on a large scale; and it is in regions of this type that I assume the first carbon material of an organic chemical nature could have been concentrated to a considerable degree. A similar situation can occur when a lagoon is isolated by elevation of the land, again with subsequent concentration of the water solution by evaporation.

We have parallels to these processes in the more recent formation of petroleum in localized regions, by isolation of a greater or lesser body of water containing a certain concentration of organic carbon in the form of organisms of all kinds. In many cases this has occurred in inland seas or lagoons, where finally the increasingly concentrated organic carbon material accumulated in extensive layers and was later embedded in sediment and exposed to geologic processes on a large scale, with subsequent changes in temperature and pressure—all leading to chemical reactions between the components of the material. The result in this special case is petroleum.

So far as I can see, we do not have to make any more fanciful assumptions for large-scale concentration processes in the geologic epoch four billion years ago. We can assume with certainty that land elevation and depression, isolation of inland seas, evaporation, and embedding and pressure treatment of concentrated or organic material must have occurred in very early times. Some rock-foldings can be dated with certainty back to about two billion years ago. There is some reason to assume that the earth's crust underwent periods of unrest even earlier than this far-distant time, when local pockets must have been formed. The early carbon material of more or less complicated organic compounds was thereby gradually concentrated in considerable amounts in a number of regions. Today we can see the result of the evaporation of seas later than the Cambrian period, resulting in petroleum. The same type of process in pre-Cambrian times, with a mixture of simpler and more complicated organic compounds, must have led to substantially the same result that occurred to organisms of later epochs.

Let us analyze the material in the sea and note how it is changed by evaporation in different localities. In sea water we can definitely reckon with a number of simple acids emanating from cyanogen and hydrogen cyanide and from certain unsaturated hydrocarbons, acetylene and related compounds—in other words, oxalic acid, formic acid, acetic acid; probably propionic acid and higher fatty acids in slight amounts; and also different kinds of organic cyanides of the nitrile type and a rich assortment of amino acids. Alcohols, both methyl alcohol and our more attractive ethyl alcohol, are represented as well as some aldehydes, but few cyclic carbon compounds. The inventory should look like the list given on page 105.

All this represents, so to speak, the inevitable consequences of the first billion years' conversion of the primary carbon material on our earth. The end products in this scheme are exceedingly stable in their own way and sluggishly reactive, with the possible exception of acetaldehyde. Here we need an incitement to activation of this material, which, to be sure, has the potential to function as the starting material for something else more complicated, but which at this stage has certain possibilities of existing as such for aeons ahead, unaffected by time, comfortably inert.

In general, the period immediately following the formation of the seas is a long period of waiting for something new. Chemical equilibria have become more or less "frozen"; the majority of the carbon material lies dissolved in the seas in exceedingly minute concentrations as non-carbonate carbon. Some activity is still taking place in

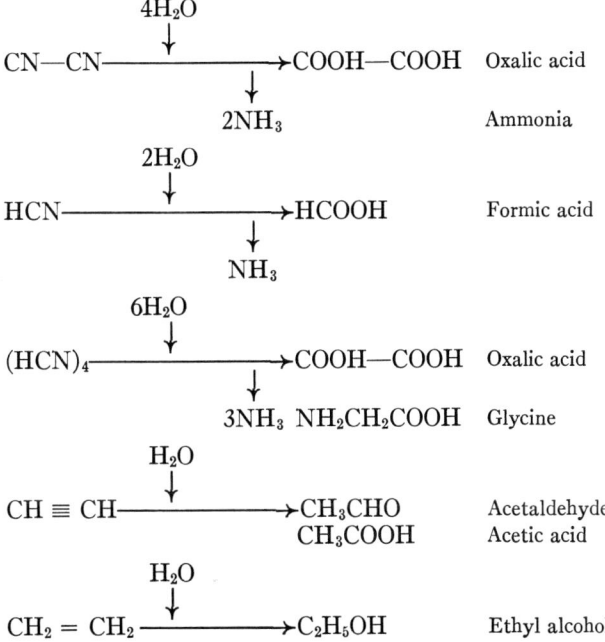

the upper layers of the atmosphere, where ultraviolet radiation continues to excite some sluggishly reactive molecules to chemical activity. Water is slowly split into hydrogen and oxygen; the hydrogen escapes, but the oxygen remains. So far this process does not take place on a large scale. Remaining methane and other atmospheric hydrocarbons are still converted on a small scale into amino acids and some more complicated compounds, which gradually are carried down with the rain toward land and sea. The earth's surface slowly and unnoticeably changes its morphology. Local evaporation takes place here and there. Sometimes accumulated carbon material happens to land in regions of volcanic activity, with thermal decomposition and simplification as a result, thus making now and then a small addition to the atmosphere. It all gives the impression of a fairly peaceful phase, chemically. What takes place, takes place quietly just now.

CHAPTER IX
COMPLICATIONS

We find ourselves now in the period one to two billion years after the formation of the earth. The environment has become stabilized into something that seems familiar. We see mountain formations, folding processes taking place, and some seismic activity, as well as rivers and lake systems, land regions undergoing erosion, transport of material, and pronounced thermal variations in different regions: tropical heat here, snowfall in other places. Rain, winds, sunshine, drought. Older layers of sediment are lifted here and there, exposing carbon material to reactions, the basic theme of which is now developed in a new environment.

The carbon compounds that are exposed are probably quite unrecognizable. They have been buried a long time in some places, and local pressure and often local temperature elevations have produced drastic transformations of the originally deposited material. In general there has been a certain degree of carbonization, the result of a gradual splitting-off of water from different combinations of carbon compounds, with the formation of more complicated material that is relatively insoluble in water—larger molecules from smaller ones, leading to polymers of all kinds.

In principle, a polymer is a chemical combination in which a certain fundamental unit, a certain structure, recurs several times within the framework of the whole molecular architecture. Our present proteins, for example, are polymers; the fundamental units have the general structure

$$-\text{NH}-\underset{R}{\text{CH}}-\text{CO}-$$

repeated, forming chains:

$$-\text{NH}-\underset{R}{\text{CH}}-\text{CO}-\text{NH}-\underset{R}{\text{CH}}-\text{CO}\ldots,$$

often with several hundred units as links. The symbol R denotes a number of variable configurations such as H, CH_3, CH_2OH, $CH(CH_3)_2$, CH_2CH_2COOH, $CH_2C_6H_5$, etc. According to very strict nomenclature, a polymer should actually contain a number of units of exactly *the same configuration;* proteins contain *the same theme,*

$$-\text{NH}-\underset{R}{\text{CH}}-\text{CO}-,$$

with individual variations, and should really be characterized as *polycondensates.* Here in the beginning we sacrifice preciseness for the sake of simplicity and collect all compounds with some repetitive group under the designation "polymers" (cf. p. 43).

This brief introduction to the concepts of polymers and polycondensates is for a further purpose than clarification of a chemical structural concept. These substances, exposed for the first time in the early terrestrial environment, constitute a pervading theme in later cells and organisms of all kinds. The complicated molecular types of cellular structure of a much later period—proteins, nucleic acids, carbohydrates in intricate patterns, polyphosphates—all are different variations of the polymerization and polycondensation theme. We meet the theme in archaic times, and we shall follow its development and possibilities.

Let us glance at the environment that for a long time forward will be the domicile of the polymers in their complicated developmental history. There are still a number of localities on the earth where violence and turbulence reign—volcanism on a larger or smaller scale, regions of heat and destruction—antipodes of the relatively peaceful chemical activity in all its forms elsewhere in the landscape. Here and there in the boundary region between the very hot and the temperate are optimum conditions for the formation of polymeric material through simple heat treatment. As an example from inorganic chemistry, we can recall the formation of polymeric phosphates from simple phosphates by heating for a certain time; one molecule of water is split off for each phosphate residue that is coupled into the chain pattern. In extremely simplified form we can symbolize the whole process in this manner:

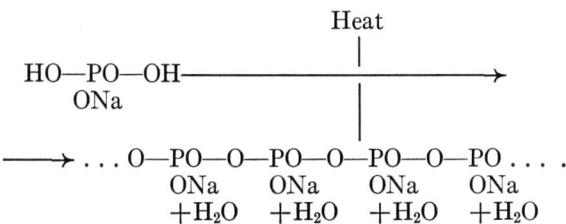

A certain simple type of phosphate is thus converted into a polymerized, so-called metaphosphate, a special variant of which with six units we use as a detergent.

This is a simple example of how thermal energy can create complexity out of simplicity. As a general rule, energy must be supplied in one form or another to bring about, by the splitting-off of water, a coupling of molecular structures into a more complicated aggregate. In the same way, by heat treatment, we can produce from simple amino acids relatively long chain structures, peptides, which bear a primitive resemblance to proteins of different kinds:

$$\text{NH}_2\text{—CH—COOH} \quad \ldots\ldots\ldots \xrightarrow{\text{Heat}}$$
$$\text{R}$$
$$\ldots \text{NH—CH—CO—NH—CH—CO—NH—CH—CO} \ldots$$
$$\text{R} \quad +\text{H}_2\text{O} \quad \text{R} \quad +\text{H}_2\text{O} \quad \text{R}$$

Oparin pointed out in the 1930's in his book *The Origin of Life* that the formation of peptide chains by heat treatment of simple amino acids probably took place at a very early stage in the history of the earth. So far as I can see, such a process is quite plausible in particular instances where a depot of evaporated material with amino acids as the dominating component coincided with a region of volcanic heat—not too far away, not too close, but just at the right distance. This means that the process could not have been one of the more common features in the chemical interplay at that time, but, given the duration of the time interval—one or two billion years—there was a relatively good chance of a slow accumulation of peptides on the earth.

Processes of this kind—the formation of polymeric structures by the splitting-off of water at temperatures between 100° C. and 300° C.—must be assumed to be the principal activity in the slow conversion of the local concentrations of material that happened to be in the vicinity of erupting volcanism. Another variant of the theme is thermal dehydrogenation of certain hydrocarbons with subsequent spontaneous polymerization, according to the fundamental scheme:

$$\underset{H_2}{\overset{R}{CH_2}-\overset{}{CH_3}\downarrow} \rightarrow \overset{R}{CH}=CH_2 \rightarrow \overset{R}{CH_2}-CH_2-\overset{R}{CH_2}-CH_2 \ldots$$

$$\underset{H_2}{\overset{R}{CH_2}-CH=CH-\overset{}{CH_3}\downarrow} \rightarrow \overset{R}{CH}=CH-CH=CH_2 \ldots \rightarrow$$

$$\rightarrow \overset{R}{CH_2}-CH=CH-CH_2-\overset{R}{CH_2}-CH=CH-CH_2 \ldots$$

The above primitive symbolization is analogous to our present-day fabrication of polyethylene and rubber material.

In addition, it is possible that formaldehyde, CH_2O, which during earlier epochs probably occurred fairly regularly during evaporation processes, could under further heat treatment polymerize itself into structures similar to present-day sugar types:

$$HCHO \longrightarrow \underset{OH}{HCH}-\underset{OH}{CH}-\underset{OH}{CH}-\underset{OH}{CH}-\underset{OH}{CH}-\underset{O}{CH} \ .$$

This reaction was studied as early as 1890 by E. Fischer in Germany and leads to, among other products, the formation of hexose fructose. The reaction is plausible in a weakly alkaline environment of temperatures that do not rapidly exceed 100° C.; such conditions could have existed here and there in the depots of organic material in early times. Formaldehyde has, moreover, an innate tendency toward condensation with a great many substances. In this type of reaction phenols become bakelite, urea becomes carbamide resins, variants of plastic products in general, complicated chain patterns of great stability, with the ability to survive almost anything.

As a general summary at this stage, we can state that the material which occasionally came to light during the epoch three to four billion years before our time contained certain general types of polymers which we recognize as analogous to the organic material of our time: peptides, quite simple and undifferentiated, but nevertheless of principally the same structural pattern as those constituting the prevailing architectural elements in our present-day proteins. There were also long hydrocarbon structures, some probably oxidized at one end—precursors to later fatty acids of longer or shorter types. Moreover, there were smaller amounts but abundant variants of primitive carbohydrates, some of which we can recognize in living organisms of today. In addition to all this, there were an assortment of polymeric phosphates and a confusing amount of plastic products of all kinds, the result of formaldehyde in uninhibited activity. In general, the material probably looked black, viscous, undifferentiated, and amorphous—what we chemists with a nod of recognition classify as the sticky mess that results in greater or lesser amounts when we try to combine one carbon compound with another.

At this point we should ponder over something we have heretofore overlooked in our analysis of the earth's earliest period. It is this feature of variability in a chemical reaction: environmental influence initiating or directing the occurrence of a new combination—a change in the environment, and the result becomes something else. For example, a reaction which runs in one direction at a certain temperature runs in another direction at another temperature. What occurs in a gaseous cloud may take an entirely different direction in an aqueous solution. What takes place in a dilute aqueous solution is different from what occurs when the material is concentrated or involved in reactions in an anhydrous milieu. Carbon compounds *do have* the general tendency in their conversion reactions of being influenced by the environment. The simplest reaction in the laboratory or on an industrial scale never results in a 100 per cent yield. A + B is never equal to AB. We always run into larger or smaller amounts of what we so elegantly designate as by-products —in moments of disappointed desperation, called by more expressive epithets. The organic material of earlier epochs was, in general, the result of what we would call derailed reactions: the inevitable result of a mass of hard-to-define material over a long period becoming complicated into something still more hard to define. From a dilute solution of something, something else was, inevitably, formed: polymeric by-products of life which gradually dominate the process as a whole.

The asphalt-like material that for long periods has been exposed to weather and wind, the legacy of transformations in darkness and under pressure in past epochs, has obviously a variable composition, depending on the milieu in which it was embedded. Bedrock minerals have had a catalytic effect on certain reactions, different for each depot. When the material comes to light, rapid differentiation takes place. The relatively insoluble is separated from the soluble; local accumulations of water will contain a relatively high concentration of organic acids, soluble carbohydrates, less of peptides. Polymeric peptide structures swim around as sediment happily mixed with drops of oil, which represent a confusing mixture of hydrocarbon structures with a certain amount of higher fatty acids. Some of the material is rapidly transported toward unknown fates, down to the sea for further evaporation and a new waiting period in limited depots. Some is filtered through sand and clay and begins to form a series of layers, with certain components collected on the surface of the filters, analogous to what we call in our day chromatographic separation. Certain types of clay gradually become impregnated in layers with peptides and other more water-soluble polymers, a combination with possibilities. There now occurs variable interaction between carbon compounds and the inorganic minerals of the bedrock. It is a separatory process on a large scale, and certain carbon compounds for some time ahead will form a symbiosis of the most primitive sort with inorganic crystal types, an idea that has been proposed by the English investigator Bernal. We assume that this combination is what is found in later stages in an organized form as envelopes, shells, skeletons. The tendency of certain polymers in an aqueous environment to form hydrogenated units begins to flourish. Gelatinous coatings on the shore here and there begin to be common features in inland lakes and smaller pools. An occasional undifferentiated lump of material floats peacefully in the clear water, saturated with salts. Some have a tendency to adsorb onto their structure certain components of the existing low-molecular material. In certain localities we find one type; in others, another. We see hydrocarbon drops emulsified in water accumulations by rain and wind, here and there incorporated in the surrounding gel. We see the main part of the material undergoing a continuous destruction by mechanical action, dissolution, and transport to the sea, to be concentrated again and differentiated in inland lake systems after existing for aeons in unreactive dilution. We see the amorphous polymers separated from the simple soluble ones displaying some activity—activity of the most primitive kind, the aim of which is to exist, to adsorb certain substances, gradually to be destroyed as a structure by mechanical forces, its own entire content and adsorbed components embedded in sediment, to come to light again only later and participate in the interplay of increasingly advanced rounds.

CHAPTER X
ACTIVITY

At the end of the second billion years polymeric material in a more or less differentiated form is probably fairly widely distributed generally here and there on the surface of the earth. The opposite poles between which the flux of terrestrial carbon compounds moves in continuous reconversion are still purely thermal: local destruction and simplification in isolated regions of volcanic heat, local concentration and complication in certain deposits under the influence of relatively moderate temperatures. Occasionally the products are exposed, the more insoluble separated from the readily soluble, during brief moments in their polymeric existence in an unhurried world; what takes place does so in geologic tempo: a largo compared with other earlier epochs of pronounced allegro. An inventory of the general environment of different regions of the earth's surface at that time reveals that both the atmosphere and bedrock are undergoing slow processes of differentiation. The composition of the atmosphere is probably dominated by nitrogen and water vapor, with smaller amounts of carbon dioxide, hydrocarbons, ammonia, and hydrogen sulphide—in short, what we might characterize as a *reducing* atmosphere. Furthermore, there is an incipient concentration of the inert gas argon, which is the result of continuous degradation of the potassium isotope K^{40} to A^{40}. At the same time, however, in the upper atmospheric layer there is constant splitting of water vapor into hydrogen and oxygen under the influence of the ultraviolet fraction of the solar radiation. This process is not very extensive, but it can be counted on to give a characteristic result in the long run. This photolytic cleavage of water in the higher layers must have released a good deal of hydrogen from the planet, because of its slight molecular weight. Oxygen remains and slowly begins to be concentrated; here and there in the boundary regions between atmosphere and bedrock certain chemical reactions are taking place, oxidative reactions.

Under the conditions that prevailed during the first billions of years, the bedrock material, exposed to a reducing atmosphere, must have existed on the lowest oxidative level. With the continuous formation of oxygen—even on the most modest scale—some minerals in the surface must have gradually been oxidized up to a higher oxidative level, thereby keeping the oxygen content of the atmosphere down to, by way of conjecture, a couple of hundredths of one per cent. There is reason to believe that the bedrock components of iron compounds slowly but surely underwent a steady transformation process, iron ions on a lower oxidative level being oxidized to a higher; in chemical language: from Fe^{++} to Fe^{+++}. At the end of the second billion years this tendency must have become pronounced, with certain consequences for the carbon cycle as a whole.

At the same time there occurs a slow but pronounced change in the oxidative status of sulphur compounds of all kinds. We can still reckon with a remarkable leakage of sulphur compounds from inside the earth, localized in volcanic tracts like those of today. The composition of volcanic gases, then as now, probably must have been dominated by water vapor and carbon dioxide, with smaller amounts of hydrogen chloride and hydrogen fluoride as well as hydrogen sulphide and certain other sulphur compounds. In some rare cases, a local confrontation of metallic oxides and metallic sulphides in the hotter interior of the earth must have led to an oxidative process whereby the metal was separated and elementary sulphur was given some chance of being exposed as such on the earth's surface via volcanization:

$$2MeO + MeS \longrightarrow 3Me + SO_2$$

$$SO_2 + 2H_2S \longrightarrow 3S + 2H_2O \ .$$

The slow accumulation of oxygen in the atmosphere similarly must have gradually had a certain influence on the hydrogen sulphide during the continuous formation of sulphur plus water—to be sure, not on any larger scale, but, with the passage of time, smaller depots of elementary sulphur and a confusing flora of sulphur compounds of different degrees of oxidation must have resulted.

As a general consequence for the carbon compounds, we can reckon with a slow concentration of sulphur in depots of complex carbon material leading to dehydrogenation processes here and there in localities of higher temperature, where the tendency of sulphur to form hydrogen sulphide with hydrocarbon structures must have come to the fore. This can be simply expressed as: $XH_2 + S \rightarrow X + H_2S$.

The result is a re-formation of hydrogen sulphide from sulphur, which in its turn was formed by the oxidation of hydrogen sulphide. At the same time, under other external conditions, a greater content of hydrogen sulphide in embedded depots of carbon material probably could have resulted in a process with a tendency toward *hydrogenation* of the material during the formation of sulphur: $X + H_2S \rightarrow XH_2 + S$.

What we see here as two counteracting tendencies is a reflex of the general differentiation process; slowly but surely it begins to give the interplay of destruction and complication a new character. Step by step the opposite poles, hitherto predominantly thermal, begin to be supplanted by a *chemical* differential representing the span between dehydrogenation and rehydrogenation. In both processes hydrogen sulphide and sulphur as well as Fe^{++} and Fe^{+++} will function as the extremes in a chemical interplay, out of which, so far as the carbon compounds are concerned, the active will gradually emerge from the passive.

The picture we can imagine of local chemical processes about two billion years after the formation of the earth is naturally exceedingly vague. We have attempted, nevertheless, to develop a fairly realistic background environment in which polymeric material existed in archaic times. Each isolated locality where polymeric material was exposed must have had certain individual features, from a chemical point of view. In some regions the concentration of hydrogen sulphide must have been high, in others low. In some places the balance between iron of a low degree of oxidation and iron of a high degree must have been displaced in the direction of Fe^{++}, in others toward Fe^{+++}. The composition of the polymeric material from innumerable constituents must have varied from locality to locality; the variable inorganic composition of the environment must have given rise to all kinds of separation processes that resulted in individual forms. We see a prevailing tendency to differentiation, from which gradually many tens of thousands of species of carbon compounds were formed, high molecular and low molecular, some in larger quantities, others in smaller, some chemically passive, others aggressively reactive.

A general feature which can be discerned here is the tendency of the sedimentary polymeric material to incorporate certain types of low molecular material. In some cases chemically active components in local aqueous solutions can be incorporated by direct chemical linkage; in other cases, certain types of inorganic substances—from simple metallic ions to more complex material—and simple carbon compounds are incorporated into the more complicated by *adsorption*. In isolated cases the latter process leads to *the formation of polymeric material with catalytic properties*.

As is well known, catalysis is the ability of certain substances or complexes to convert material without being altered themselves; to participate in, to activate, a certain tendency without undergoing any change themselves; to be active by virtue of their existence. We can recall the structure of the enzymes that in the present day catalyze innumerable chemical reactions in cells of different types. They are specifically set for certain types of chemical reactions, ready to activate a slumbering tendency to chemical transformation by lowering the threshold that dams up the potential energy content of a substance, allowing these possibilities to blossom out into activity and into new stabilization as something else.

In certain cases catalytic activity can be effected by simple incorporation of a certain kind of metallic ion into a special chemical setting. Our early polymers had definite possibilities in this direction, particularly the specialized types built as sequences of amino acids

ACTIVITY

—in other words, peptides. Other types of catalyzers have a more complicated history behind them. We can assume that at times carbon material, carbonized to an extreme degree during early epochs, landed in regions with an unusually high oxidative power. The result was an oxidation of certain modifications of the carbon substance, such as graphite forming carboxylic acids, still containing the graphite structure of carbon atoms in hexagonal patterns.

We meet these stray derivatives of organic chemistry now and then in our present bedrock in the form of aluminum salts. The technical name of the mineral is honey-stone. We know from our experience in modern organic chemistry that the corresponding acid, the fundamental substance, is a benzol-hexacarboxylic acid, and that on heating, many of the COOH-groups in the ring system are removed with the formation of CO_2, carbon dioxide. A very small fraction of this heated material contains what we technically call phthalic acid: the original material minus 4 CO_2 and with two COOH-groups immediately adjacent to each other. It is not much of the original whole; the main part becomes something entirely different, but sufficient in an environment where ammonia happens to be the dominating component to form an ammonium salt, with possibilities by the splitting off of water of forming a benzol-dicyanide. This in its turn at higher temperatures has possibilities, in the presence of certain metallic ions, of forming complexes of four 2-ring structures around a metal atom, a symmetric configuration that has great catalytic possibilities and, moreover, a thermal stability enabling the compound to survive at temperatures between 100° C. and 400° C. Seen as a whole, these substances, *phthalocyanines*, are closely related to enzyme activators of our time, hemins; their formation in archaic times can be reckoned with because of their extreme stability in all respects. Their adsorption in early polymers is a definite possibility. (See Fig. 23.)

It is clearly possible that a series of similar structures is formed from the polymeric hydrogen cyanide, plus the simple compound glyoxal, CHO-CHO, by subsequent heat treatment with a suitable metallic ion. These compounds are very similar in organization to phthalocyanines and hemins, and their catalytic properties are related. The chances that they were formed in the remote period under discussion are very good, and again we have a small but aggressively adsorbable material to incorporate into our polymers of all kinds, again leading to catalytic activity—in short, activity.

An isolated crater lake, one among many vestiges of earlier volcanic activity in the more turbulent stage, is now a relatively peaceful region; only a certain amount of escaping hydrogen sulphide is reminiscent of former days of intensity and power. The shores are resplendent with sparkling minerals in all colors, rather like present-

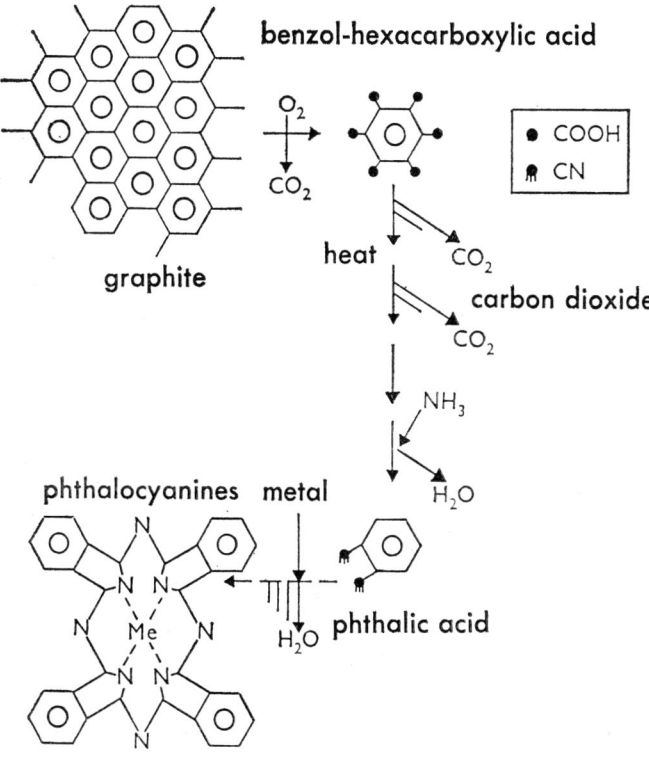

Fig. 23.—Illustration of the possibility of forming by oxidation of graphite a catalytically active compound resembling porphyrins. The yield is certainly exceedingly small; the main part of the material becomes something entirely different.

day New Zealand with its thermal springs and their display of chemical activity and iridescence. The brownish-green water harbors complex sedimentary carbon material of the most different kinds, synthesizing and decomposing, some catalytically active, others more passive, uninvolved in future calamities. The water solution contains innumerable carbon compounds of varying degrees of complexity. Minerals along the shore make a continuous contribution to a localized transformation of peripheral material. During each million years that have passed, the sedimentary slimy deposit has become more chemically aggressive: *the influence of the environment on the polymers is beginning to be reciprocated by the influence of the polymers on the environment.*

This is our general impression. As we attempt to summarize it in words, we can note a tendency: the center of chemical activity, of transformations of all kinds, has slowly passed from the more inorganic regions of the environment to the organic units that are beginning to be the framework of the carbon compounds. We perceive that the chemical process that has led to the formation of complex carbon material is now beginning to be influenced, and directed, by the material itself; the actors are taking over the stage management. What has been slowly activated to chemical catalytic aggressiveness is beginning to influence a sluggish and traditional environment. A revolution, or perhaps rather an evolution, has begun.

When we attempt to free ourselves from the more subjective in the situation, fascinating in all its impenetrableness, we find in our analysis something factual, something concrete to hold on to: namely, each locality, each isolated body of water, in itself is or functions as what we in later times designate an organism. There is a certain influx of low molecular substrate, simpler carbon compounds liberated from the polymeric complex, and a chemical differential from hydrogen sulphide to sulphur, another from Fe^{++} to Fe^{+++}. Within this environmental differentiation the simple activity of the polymers functions as enzymes within a cell. In some cases we encounter the fundamental sequence

ACTIVITY

$$X \to A \to |-B \to |-C \to |-D \to |\to F-|\to G-|\to H-|\to I \to A$$

with rate constants $_pE_1, _pE_2, _pE_3, _pE_4, _pE_5, _pE_6, _pE_7$ over the arrows, and CO_2 released below G and H.

where under the influence of different catalytically active polymers a certain compound A is continuously re-formed; it, in turn, activates the incorporation of a certain substance X into something new in combination with a link in the chain. During further transformations, one or two molecules of carbon dioxide are released from the sequence and the result becomes the substance A, ready for a new round. In some cases hydrogen is split off as a by-product: the result of a dehydrogenation in one or another step, a pair of hydrogen atoms that is rapidly accepted by another catalytic unit specialized for this type of reaction.

What we see take place within these archaic bodies of water is fundamentally what occurs in our time within the limits of a living cell. Carbon compounds of different types are accumulated in the juice from polymeric material exposed to water, all substrates for the polymeric, now with varying strength of catalytic action. The content of hydrogen sulphide varies in some regions; the content of Fe^{++} compared to Fe^{+++} in others gives the action its special character. In some lakes especially dehydrogenating reactions take place, in others hydrogenating. A specialization is in progress that in an unfathomable manner leads toward polarization: excessive hydrogenation and excessive dehydrogenation of the simple. The complicated retains its uniqueness, influenced in one direction or another depending on the environment, on the relationship $H_2S:S$, on the relationship $Fe^{++}:Fe^{+++}$.

To all this we can add something else; namely, that out of all this material, which is slowly being transformed into something new, certain substances themselves become components of the active polymers. We discern an incipient growth process, the active growing at the expense of the passive, the high molecular gradually incorporating material that results from its own catalytic action on the simple carbon compounds of the environment. At the same time we have noted another tendency: that some material begins to disappear from the scene by catalytic splitting-off of carbon dioxide, enriching the sea and the atmosphere. This tendency will gradually reduce the number of the simpler organic compounds in the environment around us: carbon dioxide and carbonates will constitute the main part of the terrestrial carbon material. Parallel to this development and accelerated by the first primitive catalysts, we see, however, the beginning of an evolution of the more advanced organizations that out of still simpler starting material will be able to create complexities. The terminal phase in this direction of development gradually becomes the utilization of carbon dioxide as such for the formation of highly organized catalytic aggregates, somewhat similar to enzymes and enzyme systems.

So far this development is restricted to the localities where hydrogen sulphide and iron salts are the motive power for the first primitive and unorganized catalytic dynamics in isolated crater lakes. From these rather unorganized systems there gradually appears a system with the ability of freeing itself from the local conditions and spreading over the water courses of the surface of the earth as something completely new under the sun. It is only two or three billion years to that time.

Out of the increasingly carbon-dioxide-enriched environment, a system develops that gradually will attain such a degree of chemical aggressiveness that carbon dioxide itself will become the substrate. The lakes exist for only short periods. The developmental tendency is to something more advanced. The more advanced becomes differentiated, activated, exists for a short time in order to activate and nourish others during its own decay. Does the situation sound familiar?

CHAPTER XI
VITALITY

The first glimpse we get of a functioning system that is reminiscent of an organism is about two billion years before our time, three billion years after the birth of the earth as a planet. At first glance, it appears to have very little in common with organisms of today. It is most like a small tarn where something resembling fermentation is taking place. In the beginning it is possible that we cannot distinguish between organism and organelles, the components of the whole. The organelles are in the form of polymeric material with catalytic activity floating around, finely suspended in the water, all simply formed enzyme components with one or another specialized function. What gives the whole the character of something organic is a kind of interaction between the organelles within the framework of the whole. Hydrogen sulphide bubbles up here and there through the brown water, colored by salts of all kinds, muddied by polymers and free sulphur in fine distribution. Iron salts are freed from the minerals along the shore by the action of the acid water. Simpler carbon compounds in solution pass between the multitude of transformation centers, some more and more hydrogenated, others more and more dehydrogenated, occasionally splitting off carbon dioxide, some added to the polymers as a link in a continuous growth process. There is a constant supply of material for treatment—for separation, solution, adsorption; and the separated material undergoes the same phases and has a strong catalytic effect on itself and the environment in one direction or another. Life? Beyond a doubt, yes!

The period that now follows I should like to characterize as a period of concentration and separatism. As I see it—naturally more as a vision than as something concrete; who can really do anything more?—all this activity in the archaic lakes and tarns must gradually be concentrated to smaller units, as if one drew a veil, a membrane, around a certain small volume in the water mass and permitted the activity on a large scale to continue in a delimited region. The phenomenon as such is not at all remarkable; in its simplest form it occurs as a natural consequence of the situations in which a concentrated solution of something penetrates another solution and the confrontation between the components results in precipitation of a relatively insoluble product, at times as a thin veil of precipitated material, at times as a thicker fabric, a membrane surrounding the scene of action. Something of this sort now and then limits the chemical activity within our archaic lakes and tarns; the great whole gradually becomes *a summation of the activity in innumerable smaller units*. At first there is a rather uncritical accumulation of enzymatic activities. A later development will be selection of the most stable and functional systems.

There is one thing we must remember regarding the development in the tarns during this period; namely, we can by no means count on undisturbed development in a locality for more than one or two hundred years. In other words, even if the whole developmental period for the selection of polymers for enzymes, enzyme complexes, organelles, and primitive organisms can last for one or two billion years, the *actual* time for each dramatic change of status must be very short. Momentary flashes of activity must be followed by passive existence in a succession of geologic strata, a hundred years followed by a couple of million in tranquillity and a slow transformation in the depths of the depots. But one thing certain is that for each exposure of increasingly complicated carbon material, particular groups of substances will reappear, especially some metal complexes with catalytic properties. As I view the whole situation, we must not overlook the fact that crystallized minerals intimately combined with polymeric carbon compounds have had a definitely stabilizing effect on aggregates of different types.

As a final stage in this long development from catalytically active bodies of water to smaller units with multiple chemical potentialities, I see a selective formation of certain aggregates with the ability of freeing themselves from their environment, the development of such a high degree of versatility that the participating units are no longer entirely dependent on *all* the components of the environment. After the whole has evolved a complex activity in its parts, the parts begin to react on the environment in an exceedingly subtle manner.

In other words, the complexity that earlier distinguished the environment as such in regard to the amounts and variety of chemical components now begins to glide over into a complexity of the aggregate in a relatively simple environment. We can reckon that at the end of the third billion years simpler carbon compounds in solution begin to become a rarity, with one exception: carbon dioxide. During the whole time that carbon material in general was being shuttled around in the layers and regions of the earth's crust, a slow cleavage of carbon dioxide from the material must have taken place, a process that in some respects can be regarded as a hydrogenation per se, provided that one defines the term broadly:

$$R-COOH \longrightarrow RH + CO_2 .$$

We thus lose some of the material as carbon dioxide but gain in the degree of hydrogenation of what remains. We familiarized ourselves earlier (page 78) with another reaction scheme whereby carbon dioxide was formed from simple material by a series of dehydrogenation reactions. These processes in combination must gradually have led to a considerable concentration of carbon dioxide in the sea and atmosphere. The developmental phase that will now succeed the foregoing will have as the sustaining feature incorporation of carbon dioxide as *substrate*, as nourishment.

We know from our experience of present-day sulphur bacteria that these primitive organisms can in certain cases survive on the spartan diet of carbon dioxide, water, ammonia, and hydrogen sulphide. The motive force in the whole process is the tendency of hydrogen sulphide in certain catalytic systems to release its hydrogen and form sulphur. The hydrogen is captured by the hydrogen acceptors of the organic material, carbon dioxide by carbon-dioxide acceptors. The interaction between these leads to stabilization of the carbon dioxide addition products, which later can become starting material for still more complicated compounds.

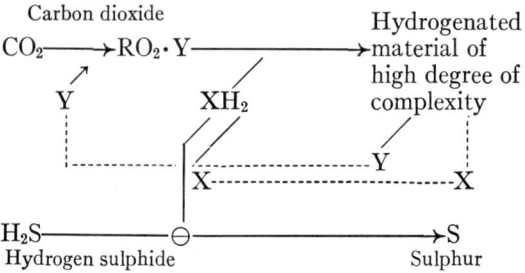

In principle, this process is a distinct parallel to that which is the basis for our present-day photosynthetic fixation of carbon dioxide (Fig. 2, page 16). It is, of course, possible that this reaction, broadly speaking, could have taken place at a very early stage in an environment with a high content of both carbon dioxide and hydrogen sulphide; but, on the other hand, it is a rather advanced process, which requires an intricate organization of enzymes in order to function. Probably this reaction is the only possible one that must come sooner or later out of pure necessity; all other substrate except CO_2 begins to disappear! And we know that the first fixation of carbon dioxide sooner or later must occur, and that the first phase of this process probably took place under anaerobic conditions: during the period when the atmosphere was still composed mainly of nitrogen and increasingly enriched with carbon dioxide, when hydrogen sulphide was still a relatively common by-product of volcanic leakage, and when the oxidized forms of sulphur SO_3^{--}, SO_4^{--} were in the minority.

At the same time we know of present-day organisms that utilize the system Fe^{++}—Fe^{+++} as the motive power for the fixation of carbon dioxide. We call them iron bacteria, with *Leptotrix* as a typical component. The activity here is based on the fact that iron at a lower oxidative stage, Fe^{++}, can in a suitable configuration have a hydrogenating effect per se:

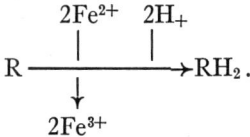

We see the quiet activity of these organisms now and then in our swamps, where they industriously but slowly reproduce and accumulate around themselves a respectable amount of ferric hydroxide, $Fe(OH)_3$, which we sometimes call bog ore.

Are the types of sulphur bacteria and iron bacteria that we designate *Thiotrix* and *Leptotrix* ancient relics of the advanced polymeric aggregates that we postulate as having existed in archaic times? It is not certain, of course, but the chances are quite good. These types of organism are the only one we can resort to, so to speak, when conceiving the first more advanced aggregates in anaerobic times, when carbon dioxide began to be the only simpler substrate and when the simplification of the environment and the complication of the aggregate had aggravated the situation. Moreover, we know that the primitive fixation of carbon dioxide represented by these organisms is analogous to the type that later became predominant, based on *water* as the hydrogen donor, with solar radiation as the agent for cleaving water into hydrogen-charged catalysts and free oxygen. If we draw a series of parallels, as shown on page 128, we see a development from the more localized reactions $Fe^{++} \rightarrow Fe^{+++}$ and $H_2S \rightarrow S$ to the earth-encircling reaction $H_2O \rightarrow O$! It is in this development that the first primitive organisms get a chance to leave the centers of chemical energy to which they have been bound for eons and to begin an existence using extraterrestrial motive power—solar radiation, solar energy—water as the medium, carbon dioxide as the substrate. Both these substances occur in abundance throughout the earth; both have a definite chance, in contact with aggressive organisms, to be incorporated, to increase the organic material, to be incorporated again, to activate the surroundings—to what? Life, of course!

I have made this sweeping survey in an accelerated tempo mainly, perhaps, to accentuate the course of development of the first organisms, but partly, perhaps, to emphasize that it is impossible to

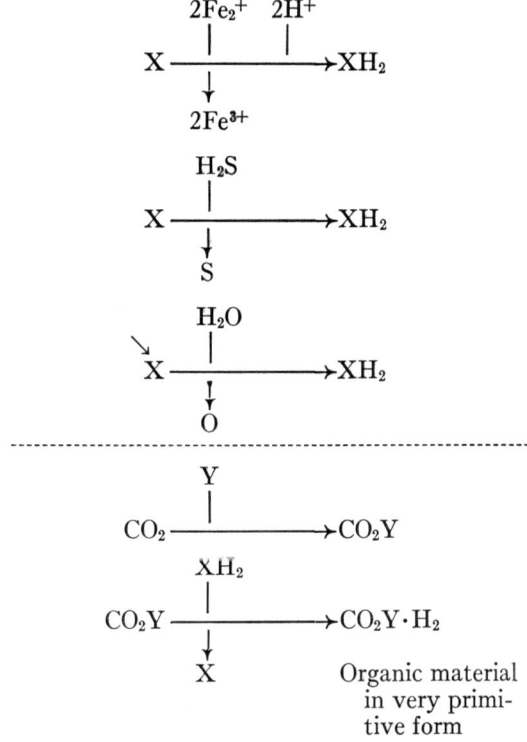

pinpoint a certain time, a certain dramatic *now*, when an organism in the present connotation of the word was differentiated from passive chemistry. The line of thought has been the following: development of polymeric, largely chemically passive material, exposed now and then in regions where there was opportunity for hydrogenation and dehydrogenation. The regions had predominant hydrogen sulphide and sulphur, or iron at lower, sometimes higher, degrees of oxidation. At times both aspects were combined, and from the whole came the idea of archaic lakes as the first representatives of chemical auto-activity, undoubtedly an extremely primitive picture of functioning organisms. Then gradually the polymeric units of the lakes took over the role of the whole and incorporated the chemical activity of the whole into smaller and still smaller aggregates. The units became increasingly complicated, the environment still simpler. Finally, the chemical complexity of the units became so great that they were able to break through the energy barrier that prevented the simple carbon dioxide from becoming the substrate for the growth of new units. At first this took place with the help of local energy sources, H_2S and Fe^{++}, later with the help of solar energy cleaving water. All this, in general, entailed a gradual change of the outline of the carbon cycle. The multitude of chemical units in the earlier environment, which, so to speak, constantly influenced primitive polymers in different directions, gradually began to become the substrate for increasingly aggressive catalytic organizations. The development progressed *from simpler polymers in a complicated environment to complex polymeric aggregates in a simple environment,* on to the epoch when carbon dioxide became the substrate and life was thereby assured for a long time forward.

CHAPTER XII
SOMETHING NEW UNDER THE SUN

The long period during which the first organisms now develop in an environment of sulphur compounds and iron salts will be an exceedingly fruitful epoch from a chemical point of view. We can reckon with many variants of the two themes, which gradually drive the development toward the first fixation of carbon dioxide with the utilization of carbon dioxide as the substrate for growth. When this fundamental process has begun to function here and there in delimited localities, life on the organismic level, so to speak, is secured, at least for a long time forward, and the primitive organismic material has a chance to develop its potentialities in different directions in relative peace and quiet.

Within this evolution of more or less prosperous sulphur-and-iron-dependent organisms, varying both morphologically and chemically, we discern certain tendencies which later will have a fundamental effect on the process as a whole.

As organismic life develops, there is a continuous *improvement of the energy metabolism* within individual units. If we consider for a moment our present-day organisms, as discussed on pages 33 ff., we can see that a prevailing feature is a mechanism for temporary storage of energy based on a complex system of labile phosphate compounds. We recall how certain conversion processes, acting on certain substances in the cellular metabolism, include incorporation of inorganic phosphate in combination with the substance in question. A later transformation, involving the splitting-off of hydrogen—or, at times, water—can cause the chemical activation of that corner of the molecule to which the phosphate is bound. Simply expressed, this means that the energy content of the molecule is concentrated in a certain small region. This region of chemical aggressiveness, localized in the bound phosphate, can further react with different substances; one variant of what can occur is a combination with a certain other type of phosphate unit and certain chemical substances, where the configuration itself of two or three phosphate residues, linked together as di- to tri-phosphate, implies a storage of aggressiveness, configurations capable of mobilizing the bound phosphate to arouse the slowly reactive molecules to activity. This activity can later be reflected in syntheses of all kinds and in the creation of cell material with its organization of enzymes.

This interplay of active phosphate compounds in our present-day cells illustrates a general principle: *by gradual destruction of the major part of the multitude of the simple, energy is concentrated to activate a minor fraction of the whole to a higher degree of complexity.* We met this principle earlier during the development of life under more drastic forms of destruction and new combinations: from the action of this mechanism on the carbon compounds *in corpore* between the opposite poles of relative heat and coolness, we gradually arrived at individual formation on the same principle within innumerable complex systems in a world of rather moderate thermal differences. We see in the earliest organisms during the sulphur-and-iron period a development of energy transference with active phosphate as the intermediary, a forerunner to present-day variants of the same interplay under considerably more complicated conditions.

If we try to imagine the nature of the active phosphate compounds during early times, our choice probably falls on the simple polymers of phosphoric acid, the poly-metaphosphate type, which have had a good chance of being formed purely chemically by thermal treatment of simple phosphates during the splitting-off of water (see page 109). Today we have in certain cells enzyme systems that can cleave metaphosphates into simpler units, thereby giving the cells a chance to utilize the energy stored in the metaphosphates. This process probably took place in the first enzymatically active aggregates. At first they utilized only the available supply of metaphosphates in their surroundings; later they developed the art of reforming the same type of compounds during enzymatic decomposition of some carbon compounds in the environment. Thus, the process becomes gradually modernized, and in time more functional and advanced phosphate compounds will participate in the individually formed energy conversions with ATP and its analogues as novitiates formed at the end of the fourth billion years.

As a natural consequence of the general development of energy conversion via active phosphate compounds, there is a slow increase in the ability to synthesize still more advanced material of varying structure. When the simplest sulphur- and iron-dependent organisms began to incorporate the interplay of the phosphate compounds into their systems, their general chemical versatility must have increased to a surprising degree. What we previously suggested as a terminal phase of this development—the ability to incorporate carbon dioxide as the substrate for individual growth—must be characterized as a direct consequence of the formation of an organized interaction whereby sulphur compounds, iron complexes, and phosphate compounds in combination act on carbon compounds of different kinds—both on the temporarily primitive organic aggregates and on the material in the environment that is the substrate for growth and maintenance of the systems in their individual forms. The ability to fix carbon dioxide is a logical consequence of this developmental tendency, the suitable catalysts being gradually developed and selected over millions of years.

As a refreshing digression from this technical discussion, which is a crude attempt to outline a chemical background for the development of the primitive organisms that we tried to reconstruct in the preceding chapter, we can try to familiarize ourselves with the conditions at the time when fixation of carbon dioxide had become established for our organisms and when development was ready for something completely new under the sun. In the locality we have chosen as observation point, we see something we have come to regard as a center of life: a crater lake whose shores are lined with deposits of sulphur and more or less differentiated organic material in the course of dissolution. In the water there is a fine alluvium resembling small gelatinous bubbles with finer bubbles covering their membranous structure of gel and inorganic precipitates of varying pattern. Hydrogen sulphide is, consequently, still present in higher or lower precipitation of sulphur along the shore and toward the bottom.

The atmosphere, increasingly enriched with carbon dioxide from the activity in general, has, over the course of millions of years, begun to influence the degree of escaping infrared radiation; possibly the temperature during certain periods has risen a couple of degrees around the earth as a whole. The sky is the same greenish-gray tone as earlier, usually covered by heavy clouds. Rain is the rule, due to the cooling effect of the high mountain chains. We see at a distance snow-capped peaks, presaging some freezing on a larger scale—which occurred now and then during the last billion years, indicating that the carbon dioxide content of the atmosphere, in spite of all the enrichment, remained at a concentration of less than 0.1 per cent. From a nearby peak we can contemplate the desolation of the landscape, the alternation of ridges toward the distant mountains, yellow and brown-red interspersed with dark lines and points, water courses transporting life and non-life. The sea is visible at a distance—for the time being the terminal station for organic life in the most advanced stage of its development. Carbon dioxide is slowly taken up in the sea water, in equilibrium with the atmosphere, and the calcium and magnesium salts accumulated in the sea are slowly confronted with a changing carbon dioxide content. Soon there comes a period of large-scale precipitation of relatively insoluble carbonates. It is only a question of time.

In our crater lake we occasionally observe unusually brilliant colors. This is the result of a very advanced chemical development that now leads to the formation of metal complexes of exceptionally high catalytic ability. Some appear to be old acquaintances, related to phthalocyanines from earlier epochs. They contain a metal atom in the center of a complex of four ring structures. We recognize *porphyrins* in these fundamental units, which have a definite similarity to present-day metal porphyrins—hemins, cytochromes. We recall their structure in all its symmetric beauty (Fig. 24).

Some of the greener substances we find to be variants of the theme: the same ring system, in the main, however, with a long hydrocarbon chain neatly incorporated into the whole. In the center is a metal atom, occasionally magnesium. We think of our present-day chlorophyll and the role it plays in the fixation of carbon dioxide. It is possible that we are on the way to a new developmental phase? Let us wait a few more million years!

The gas bubbles around green sacks that look like organized slime which grow continuously and free themselves now and then

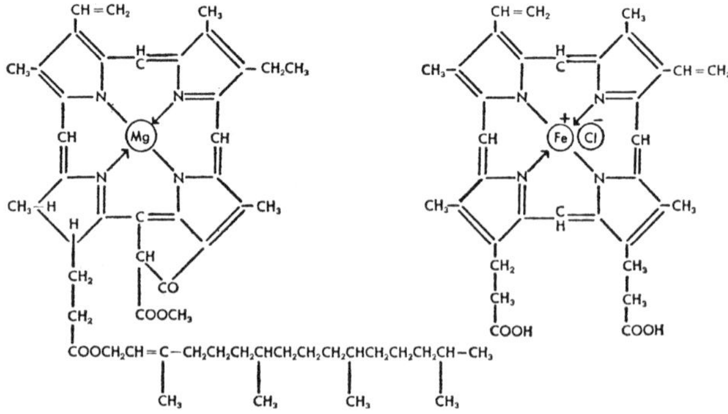

Fig. 24.—Main features in the symmetric structures that constitute the fundamental unit in what we call *porphyrins*. The side groups can be of different types, often $-CH_3$, $-CH=CH_2$, $-CH_2CH_2COOH$. In the porphyrins, which in our cells act as intermediates in the final oxidation of hydrogen-charged substances of the type XH_2 (see Fig. 8, p. 38)— thus what we call *cytochromes*—the central atom is an iron atom in different degrees of oxidation: $Fe^{++} + Fe^{+++}$ in rapid alternation. Certain porphyrins—for example, one component of hemoglobin in our blood— have similarly a central iron atom and act as an oxygen conveyor. Certain other porphyrins in nature can contain other central atoms than iron. Chlorophyll of modern green plants is a porphyrin with a central magnesium atom and with a complicated side structure attached to one site in the molecule. The porphyrins frequently occur in the mitochondria of the cells, bound to certain specific proteins. The combination acts as an enzyme. Metal porphyrines with a general structure related to that of chlorophyll have actually recently been found in fossil carbon material from the Cambrian, and even back in the pre-Cambrian, around 1.5 billion years of age. Like the phthalocyanines (Fig. 23, p. 119), porphyrines are very stable molecules.

from the organism on the surface of the water. During the time we last observed life in its regional form, we began to notice a new shade in the sky, a bluish tone, increasingly intensified over the last million years. The accumulations of sulphur along the shore have not increased; rather, they appear to be undergoing some sort of corrosion. The whole vicinity has become spotted with pronounced reddish brown; the water courses around the region carry green material to the sea, actively splitting off water, producing oxygen, fixing carbon dioxide, growing as pondweed in stagnant water, whirled up as green slime in rapids and falls. In the lagoons around the sea, increasing green masses appear; occasionally one encounters a fully vital unit on its way out toward an unknown fate. Life has undergone a revolution: the units are no longer confined to the environment of hydrogen sulphide and iron. H_2S has been replaced by H_2O, chemical cleavage of H_2S to $2H + S$ by the solar cleavage of H_2O to $2H + O$; the earth is open to life of all kinds; substrate exists everywhere; *vive la liberté!*

The first period of the development of photosynthesis on a large scale must have seemed like a wild frenzy, a carnival of liberated chemical possibilities. Oxygen must have been produced at an accelerated rate. Probably over the course of a million years, perhaps only less than 100,000 years, the composition of the atmosphere became about the same as it is in our day with approximately 20 per cent oxygen. We can reckon that green plants at present enrich the atmosphere with oxygen at a rate that—were the activity not counteracted by animalia with their corresponding oxygen consumption —would entail a doubling of the present oxygen supply in about 5,000 years. The early activity of photosynthesis must have been uninhibited and powerful during the first developmental period, since no organisms of a modern respiratory type were on the spot to consume either oxygen or the advanced newcomers producing it. However, the increasing oxygen in the atmosphere must gradually have acted as an inhibiting factor per se, for it must gradually have had a somewhat toxic effect on most of the organisms around the globe. For the majority of organisms of conservative anaerobic type, the increasing content of atmospheric oxygen must have been something of a catastrophe, a brutal intervention in their metabolism, based as it was on the cleavage of hydrogen sulphide or the utilization of iron salts on the lowest oxidative level as sources of energy. We can certainly assume a wholesale eradication of organic material, resulting in a good and nutritious environment for the few types that were capable of survival. Some of these must have been able to adapt themselves at a fairly early stage to the new conditions by developing specialized enzyme systems *suitable for utilizing oxygen for dehydrogenating complex organic material.*

At this stage we can discern the development of the first respiratory, oxygen-consuming organisms. The first phase in their formation probably involved the fact that certain metal-porphyrin structures in suitable linkages are especially adapted for capturing oxygen, thereby giving dehydrogenation reactions of all kinds a chance to function in the fundamentally simple reaction, $2XH_2 + O_2 \longrightarrow 2X + 2H_2O$. We recognize this reaction from previous exposition of the metabolism of modern aerobic cells (and we know, moreover, that the above formula is the result of a long series of reaction steps). At the same time we note that this reaction, written in reverse order $(2H_2O + 2X \longrightarrow 2XH_2 + O_2)$, is the main point of photoactivated water cleavage in the first phase of photosynthesis. The first respiratory organisms that we see represent a trend, opposite to that of photosynthesis; in later phases they will be designated as Animalia, "animals." From now on they will have an effective blocking influence on the uninhibited spread of the plant world over the surface of the earth.

This development of photosynthesis as emergent environmentally dependent life contains within itself the two poles in the power play of life which we today understand as photosynthesis versus respiration. Both processes must have developed almost side by side: first photosynthesis, with enrichment of oxygen in the atmosphere as a secondary effect; later adaptation of certain organisms to this new agent during successive readjustment of their metabolism to an effective dehydrogenating mechanism—organisms capable of utilizing the surrounding complex carbon material, living and dead, for their own growth during continuous decomposition of the main part into carbon dioxide and water. At first, perhaps, two types of organisms were formed, radically opposed in type and function. Gradually the two aspects of metabolism probably become incorporated to different degrees into what later became our present-day plant world, where we meet *both* photosynthesis and respiration in complicated organization, with the former process predominant.

Regarding the dating of this explosive readjustment of metabolism of the organic material around the world, we can refer to our earlier discussion of the ratio between the sulphur isotopes S^{32} and S^{34} in sulphurous material from early times. The fairly pronounced change in the relationship between the periods prior to and after 800 million years ago—from 22.08–22.16 to 22.95–23.05 for sulphide mineral from the pre-Cambrian period compared with the same mineral from later times—has been interpreted as a transition from anaerobic sulphur metabolism in early organisms to a corresponding aerobic one; in other words, a transition from life existing in an oxygen-free environment to life in surroundings with a pronounced oxygen content. The time for the change should thus lie around 800 million years before our time, 200 to 300 million years before we meet differentiated representatives of the vegetable and animal worlds in the Cambrian seas.

The pre-Cambrian landscape has slowly taken on a new shading in the iridescent play between the atmosphere and the earth's surface. The gray-green of the sky has gradually assumed a tinge of blue, which has become more and more accentuated as the newly formed oxygen is converted into ozone in the higher layers of the atmosphere. The seascape near the shore seems familiar, with the sun and blue sky and sea contributing nuances from ultramarine to deep violet. The surface layer in constant motion contains life in intensive activity, something similar to cells and cell organizations —green, red, blue, transparent and opaque, studded with mineral crystals. Fresh oxygen is mixed in the froth of the wave crests; carbon dioxide is taken up and given off again during the short periods the participating units exist in a life which now is here to stay. From rivers and water courses more and more primitive organisms are carried toward what is to become their milieu for about 400 million years to come. The seas' riches of simple and uncomplicated food substances are used at an accelerated tempo in the interplay between countless types of living things, representing two opposing themes interwoven in intricate counterpoint.

CHAPTER XIII
RECOGNITION

A question that constantly recurs in connection with the differentiation of the most primitive units during the early epochs is: How was a stabilized succession of generations gradually formed? In any case, sooner or later we must assume that development followed a course of genetically controlled cells and cell systems, sequences of generations after generations, repeating the same themes with few variations. Everything we see around us is repetition of different forms of existence, from the simplest yeast cells and bacteria in continual division to the complicated systems for reproducing species that make a snail, a sea gull, or a human, into a snail, a sea gull, or a human.

This is a passus in the history of life which is extremely difficult to span in an acceptable manner. It is, of course, somewhat tempting to overlook this point and to contend flippantly that nature, which had so many possibilities in the past, should also be capable of giving the first more advanced organisms a suggestion concerning perpetuation. On the other hand, this very problem is so controversial that one would like to take a position, even though, naturally, one runs a risk of developing lines of thought of dubious value. This risk must be taken; one is in good company.

First of all there is the fundamental question whether our cells developed from primitive material that began with the advanced synthesis of what we now call *genes* or whether cell material more or less differentiated formed first and the genes, so to speak, entered the game at a later stage. J. B. S. Haldane believes that the simplest organism should represent a chemical structure corresponding to present-day deoxyribonucleic acid, DNA, whose ability to reproduce itself in a suitable environment gives an existing cell its pronounced degree of hereditary stability. Perhaps we should call to mind our earlier picture of the genic influence on the chemical action in living cells in general.

As previously mentioned when we discussed the fundamental chemical scheme in a cell, the protein synthesis is under the influence of different variants of ribonucleic acid, RNA, and—so far as we can see—different RNA structures influence the formation of the amino acid sequence of the proteins. RNA thus acts as a kind of executive agent; each RNA-structure of four different units repeated as a code pattern influences the formation of a pattern of some twenty amino acids, to make the characteristic shape for each individual type of protein. DNA stands above RNA and acts as a controlling agent, probably only by existing. We know that DNA is localized in different loci in the linear genic pattern of the chromosomes. Whether each locus contains one type of DNA or whether there is only one characteristic type of DNA for each cell type—and its concentration in different loci gives a feature of variability—is an open question. We assume, however, that the genes of the chromosomal structures, via DNA, in some way control the synthesis of a supply of RNA in strict repetition, and each RNA directs the synthesis of a certain definite sequence of amino acids in individual protein syntheses. The proteins that happen to be enzymes direct the flow of nutrient conversions, and RNA is one of the many cellular components which are formed from simple nutriments. After the cell has passed through the stages we call cell division, a new supply of DNA must be formed, whereupon the action proceeds again, now in two cells, and so on ad infinitum.

Thus, the cellular activity is *controlled* by DNA in the genes, *effectuated* by RNA in the synthesis of enzymes, *directed* by the enzymatic treatment of simple carbon compounds, which in turn leads to new formation of DNA, thereby of RNA, thereby of enzymes. In all this, the structure of DNA appears in some way to be the designer which makes this or that cell just that cell, that organism just that organism.

This is the situation of today, and in our attempts to clarify the origin of this self-controlling system at some point in the past, we naturally resort to the postulated properties of DNA as the controlling agent and thereby build up our theories with DNA as the starting point. On the other hand, DNA and RNA are by no means simple structures. They both are polycondensates of four different units of different ring structures with nitrogen incorporated in different patterns, linked to two types of sugar, one for DNA and one for RNA; at one end of the sugar structure there is one phosphoric acid. (Here we may recall Figure 13 on page 45.) The interlinkage may possibly be a simple problem chemically, but the constituent structures are so intricate that it is difficult to accept them as a dominating feature in the epoch when active polymers began to be differentiated into something resembling living organisms. Moreover, we know that the formation of only one single structural component of DNA or RNA in our day requires a delicate interaction among a greater number of enzymes, which gradually build up the structure from nutrient raw material. Great art on a minuscule scale.

However, it is possible to analyze the origin of hereditary stability from an entirely different starting point. When a primitive metabolic unit (let us for the sake of simplicity call this a cell even though it has not yet acquired the characteristics of modern cells) has a certain established activity pattern, therein is the basis for a very simple form of heredity. When discussing the development of life in small lakes we mentioned in passing that a later phase must have involved local isolations of the activity into a number of units, separated by some kind of membranes. We can assume that the activity as a whole proceeded regardless of whether the units were of meter dimensions or divided into a myriad of smaller units. If there was a tendency for more complicated organic material to grow from less complex, *this took place*. In this archaic situation we can assume that the growth was fairly even over the whole, each delimited unit operating about the same. At this point some factor must inevitably have disturbed the system now and then—mechanical disturbance, agitation, swirling—and smaller units must occasionally have been exposed to such stresses that they were divided, in most cases with loss of their contents, in some cases with preserved organization, thus resembling a sort of division into two or more units. Such a process gives rise to two identical organizations, both with a growth tendency. The same phenomenon repeated in many cases, in thousands of cases, is like primitive cell division. Like begets like. Actually like divides into two smaller parts, both having a chance to grow in a uniform environment into their prototype, and so on ad infinitum. What happens when there is a change in the environment?

We can make an assumption here, the legitimacy of which time must determine. It is that out of the activity of the most primitive organisms, certain substances were formed as by-products so relatively insoluble that they remained within the units and were divided fairly evenly on mechanical cell division. These substances *were* structurally the result of the activity of the whole cell and had a certain catalytic effect on some of the cellular functions. With further development along this line, we have a collection of relatively insoluble substances, all with one or another directive action on the cell as a whole. In certain respects they could be characterized as enzymes, but they did not have to be of a protein nature, and actually they could be extremely inactive from a chemical point of view. Their function is based on their mere existence, and just by existing they influence enzyme reactions in one direction or another by contact with themselves.

If we assume here that during the course of time some of these substances became generally occurring cellular products, reacting in their turn on the enzymatic functions that produced them, a sequence happens to form that appears familiar; namely, out of an activity an agent is formed that directs the activity from which the agent itself is generated, all at the expense of a remarkable mobilization of material and energy.

It is in such systems that we might conceive polymeric nucleic acids of the DNA- and RNA-type gradually being formed during such situations where environmental variations altered the activity in one direction or another. So long as the environment was fairly stable, the type of activity in isolated units must have been rather similar in different localities. A change, chemical or thermal, must have influenced divergent formations in the activity of the units. The presence of one or more relatively insoluble by-products with catalytic action on the course of the reaction must thereby have had an inhibitory effect. When RNA and DNA enter into the game, early or late, as by-products of metabolism in general—more or less undifferentiated—it is probable that their special structure was particularly adapted to act as an inhibitory agent in situations of fluctuating supply of substrate, fluctuating temperature, fluctuating supply of energy. Gradually, under the influence of DNA and RNA, chemical conservatism will emerge that is a cell is a cell is a cell. . . .

We previously suggested the possibility that the organisms which we observe today as representatives of sulphur bacteria and iron bacteria are relics from time past, before photosynthesis with solar activation entered the terrestrial arena. At the same time we assumed that these organisms at an early stage had such a high degree of chemical conservatism that they were able to maintain individuality for an enormously long time. In other words, we reckon here that a kind of genetic control existed at such an early stage that the subsequent photosynthesizing organisms had such stable individuality that they were self-sufficient in the varying environments where they were rapidly developed, freed from the earlier stable environment of H_2S and Fe^{++}.

We have made an attempt here to trace an unbroken line of thought for the gradual course that led to genetic stability in cells and organisms in general. In summary, we can say that the local environment, or, more correctly, the local environments, where the first primitive metabolic units were separated, were relatively stable for such a long period—a hundred years or so—that here and there the activity in the units formed substances with stabilizing influence on the metabolism as a whole. Gradually certain more conservative types were selected, according to the nature of the stabilizers. In some of these, during continuous destruction of the main part of the material by cessation of the activity for millions of years, there arose as by-products something like DNA and/or RNA, substances with extreme re-coupling ability in the direction of stabilization of the units in question. The final stage in such development should be the formation of what we know today as DNA and RNA, with their intricate interaction in the synthesis of enzymes and proteins in general. Concisely expressed: the stabilizers in cells today in the form of genes in chromosome bodies should have originated as *byproducts* of early chemical activity in loosely organized units of catalysts; their formation a priori, with subsequent organization of organic material into stable metabolic units of cellular nature seems —at least to me—more improbable. The correct version of this will gradually be determined. One thing is certain, it is not an easy task.

We have reached the period about 700 million years before our time. Photosynthesis is flourishing and the degree of the acidity of the seas is slowly tending toward neutral and weakly alkaline, resulting in a general tendency for carbonates of calcium and magnesium to precipitate. These and other relatively insoluble inorganic combinations begin slowly but surely to separate and to rain down, so to speak, through the clear waters, gradually forming bottom sediment of carbonates of different kinds. This tendency has some connection with the formation of crustaceous organisms, which from this time begin to become prominent in the pre-Cambrian flora and fauna.

There has been considerable discussion concerning cause and effect in the epoch when calcareous shell began to be a vogue for organisms in general. Some authorities are of the opinion that certain organisms through some genetic caprice, one mutation or another, happened to form enzyme systems capable of incorporating inorganic material into cells and cellular bodies and in so doing happened to organize shell structures, different for each case. We know that marine organisms of later periods often displayed a pronounced ability to concentrate certain substances from the dilute solution of the environment within and around themselves. It is probable that this tendency existed in organisms of all types long before the Cambrian, but that it was activated just at the period when carbonate precipitation became pronounced. In other words, a marked environmental change was involved; tendency toward carbonate precipitation in the seas in general forced organisms exposed to this environment to form inorganic structures in and around their gelatinous frailness. The carbon dioxide production of the organisms themselves probably had a certain influence on the formation of these skeletal and shell structures, which in some cases became a common covering for cellular units or organisms of different types. This tendency should result in local symbiosis of cellular units in cramped quarters, for the hereditary propensity of each cell to counteract environmental influence must have caused some conflict within the common sheath. Probably such a more or less compulsory cohabitation of certain organisms and cell types introduced a new developmental phase by virtue of the increased possibility of local exchange of metabolic products with a modifying influence on the individual units and thereby on the whole.

In this epoch of increasing shell structures—probably manifested in the beginning by the formation of more or less carelessly capsulated organisms, later by the development of admirably well-formed carbonate sheaths, later yet with the addition of chitin and other stable hydrocarbon material—we catch the first glimpse of what in later phases will represent the type of organisms we know as Cambrian algae and invertebrates. Life on the organism level has at this time acquired a feature of striking individuality; the constancy of type, morphological and functional, for each organism during many generations indicates a pronounced genetic stability, which in some cases will lead to strongly conservative lines of development for millions of years to come.

The water is tepid and the shores are swarming with life. Some units of unmistakable trilobite type are crawling around in shallower parts, and we realize with joy that we have only about 500 million years to go until our own time. Primitive organizations resembling algae are floating in the water as far as we can see, specialized for photosynthesis, blue, red, green, their colorful remains scattered over the region. Along the shore we see shells of different shapes, which in a later period will convey a message about the life and customs of certain organisms in the epoch when the development of life attained a secure position.

Around us we begin to recognize acquaintances; some of the

types we see will exist for a limited time and then leave room for other variants; some will maintain their external façade intact through innumerable generations until our own time. We recognize here a couple of these conservative types, ancestors of our present-day horseshoe crab, *Limulus*, and the little mussel-like brachiopod *Lingula*, both of which seem to have passed through the labyrinths of later evolution without having been influenced greatly in regard to type and function.

While the sun's rays fall obliquely over the dunes, we ponder the long development of the innate tendency of carbon compounds to complications, which has finally resulted in life on the organismic level, and which probably on other planets with similar external circumstances has yielded, or will yield, something like what we here take to be motion and activity without intent. The development of life seems to have a feature of inevitability, something that in the presence of suitable sources of energy blossoms forth from simple carbon material into the degree of technical perfection that is organic life in different forms. We recall a later developmental phase, when certain organisms on our earth acquire an advanced ability to comprehend time and space and thereby look back on the course of events that is their own developmental history.

As the twilight deepens, our thoughts circle around the numerous enigmatic problems that we have encountered in our confrontation with life in different developmental stages and forms, all the intricate chemical interplay that gradually developed into individual variations of a common theme: *material in continuous transformation between situations of destruction and re-synthesis, individual combinations of which were developed, within the framework of a whole, to give the drama between death and regeneration a still richer register.* We have tried to capture something of this theme in all its abstract beauty and at the same time something of what binds us together with the surrounding life in the form of cells and organisms in all their abstruse activity. Our eye absent-mindedly follows a little thing wandering along the waterfront among algae and shell remains, uncognizant of the force of the tidewater, which perhaps in a couple of hours will leave him alone in an arid world of vicissitudes. His trusting purposefulness in his wandering among all reminders of dissolution and vainness awakens certain memories of something we encountered earlier, or was it later? Where he disappears in the shadows along the shore he is a symbol of the development that has realized in the transitoriness of the individual the continuity of something greater, that is life.

NEW ORLEANS
June 19, 1960

CHAPTER XIV
COMMENTS

The lines of thought and data included in this survey of the problems concerning life and its origin are, of course, connected with the literature, both technical and popular, which has been published on the subject in recent years. There are extremely well-written and informative discourses on different parts of the problem. First of all, there is the classic work by A. I. Oparin, *The Origin of Life*, the first edition of which was published in Russian in 1924 and the second and third editions (1936, 1941) both in the original language and in English. A fourth edition is now available in English (Edinburgh and London: Oliver & Boyd, 1957) and should be regarded as a standard work with an excellent bibliography. For those who are interested in further study on the problem of the origin of life, Oparin's book is required reading—and an unusually pleasant acquaintance—even if it presupposes a rather wide knowledge of natural science in general.

From the English-speaking world there are three works which treat different aspects of the same theme in different ways. We have *The Dawn of Life*, by A. L. Rush, which gives a good survey of the relation of the problem to geochemical and astrochemical data. Furthermore, there is a very readable book by H. F. Blum, *Time's Arrow and Evolution* (Princeton, N.J.: Princeton University Press, 1951), which gives the biochemical background for the treatment of the question of the origin and development of organisms. In addition, there is a small but concentrated book by J. F. Bernal, *The Physical Basis of Life* (London: Routledge & Kegan Paul, 1951), which gives many interesting viewpoints, particularly on the interaction between organic and inorganic material in early times.

All these works treat the question of the origin of life as a problem concerning the terrestrial material's possibilities of forming complex molecular types from simpler ones, which in this survey has been called the innate tendency to complications of matter, especially of the carbon compounds. The treatment of data and the argumentation quite naturally vary considerably in the cited works. My own presentation of the same theme differs to a certain extent from earlier expositions in that I have attempted to distinguish between the concepts of life-as-a-cycle and life-as-organized-units and thereby tried to follow the development of the latter as a consequence of the formation of the former, from its first formation as a cycle in prebiotic times to the period when the same cycle in a new edition comprises interaction between organisms and the components of the inorganic environment. Of course, opinions may vary as to the value of such a presentation of the discussion. As for myself, I regard it only as a practical method for avoiding the pitfalls that everyone encounters in attempting to analyze the fundamental features in the origin of life on our earth starting from the first formation of organisms.

In addition to the many compendia concerning the origin, generation, and development of life, which in different ways have influenced my own view of the problems, there is, of course, a tremendous amount of technical literature relating to certain particular questions. It is quite impossible to give an account of all the works that during the course of twenty years have stimulated new viewpoints regarding the problem as a whole; a complete bibliography would easily fill more than a hundred closely printed pages. However, by way of guidance for those who are interested, I should like to recommend the following comprehensive works, which in turn contain valuable references to extensive original literature:

For questions concerning the origin of the planetary system:
  H. ALFVÉN: *The Origin of the Planetary System*. New York: Oxford University Press, 1954.
  H. C. UREY: *The Planets*. New Haven: Yale University Press, 1952.
For questions concerning planets in general, their development, and chemical characteristics:
  G. P. KUIPER: *The Atmosphere of the Earth and the Planets*. Chicago: University of Chicago Press, 1949.
  H. C. UREY: *On the Early Chemical History of the Earth and the Origin of Life*. Proceedings of the National Academy of Science, XXXVIII (1952), 351–63.
For questions concerning the earth's structure and geochemical data in general:
  K. RANKAMA and TH. G. SAHAMA: *Geochemistry*. Chicago: University of Chicago Press, 1952.
  V. M. GOLDSCHMIDT: *Geochemistry*. ed. A. MUIR. Oxford: Clarendon Press, 1954.
  The Earth as a Planet, Chicago: University of Chicago Press, 1956.
For questions concerning isotopes:
  K. RANKAMA: *Isotope Geology*. London: Pergamon Press, 1954.

In addition to all this there is, of course, an incredible amount of purely biochemical literature, mainly in the form of original articles but also included to some extent in compendia as course literature for special students. From the mass of respectable books in this field, J. Fruton and S. Simmonds, *General Biochemistry* (New Haven: Yale University Press, 1959), appears an excellent product; it is, however, rather difficult reading for anyone without the necessary background. On the whole, it is hard to guide a person who wishes to get a first insight into the chemical organization of life in all its varying formation. A few years ago a comprehensive work was published in Swedish, *Boken om Naturen* (Stockholm: Forum, 1953), which gives in its own way in a rather elementary form the fundamental features of the interaction between organisms and their environment and certain aspects of their origin and developmental history. In this popular science book there is a treatment of the origin of life theme that to a certain extent has a direct connection with what is presented here in a more developed form.

Concerning the disposition of this little book, *Chapter 1* should be regarded as a preparation, a kind of preliminary training, for the further treatment of the concept "life" in its different aspects. It is, so to speak, a first reconnaissance of the terrain in order to determine the contours of a theme to be elaborated further—no more, no less.

Chapter 2 contains some data concerning amounts and annual turnover of different elements within the biosphere, where naturally some figures are impaired by a certain questionableness. Particularly the distribution of organic carbon between the sea and land regions has been subjected to several revisions during recent years, as well as the distribution of carbon dioxide turnover within these regions. The order of magnitude of the stated amounts and of the turnover as a whole, however, is probably correct; complete exactitude here does not, in any case, influence the discussion concerning the form of the turnover. The gist of the section is the presentation of the chemical dualism between photosynthesizing units and the multitude of respiratory cell systems, which has been set forth here in a highly schematic form.

Further, the unique position of carbon as an element is emphasized, with its peculiarity in the matter of the variability and stability of its compounds. The latter could possibly have been stressed more strongly, in view of the discussions that occur now and then regarding the possibility that other localities in the universe should be able to maintain life on the organism level based on other elements than carbon—for example, silicon. Personally, I am extremely skeptical of such proposals, considering the limited potential which elements other than carbon possess for forming stable high molecular complexes which have at the same time versatile chemical metabolism such as represented by terrestrial cell material. When our neighboring planet Mars is investigated in detail, its organic material—if it exists—will certainly exhibit some carbon compound components, and we shall recognize some familiar structures, both low molecular and high molecular.

Chapter 3 caused me considerable trouble, since it contains a first presentation of structure and metabolism in a living cell, something which really requires ten times the space for a reasonable exposition. There has been no possibility of avoiding here a presentation in chemical symbols, which makes the whole thing difficult reading for non-chemists, even though attempts were made to express in general terms the course, the principle, of the dynamic chemical action within a cell. For those initiated in biochemistry, the chapter appears to be suspiciously simplified, and certain phases of cellular metabolism treated cursorily, to say the least—if treated at all. The aim of the discussion, however, is to give a glimpse of the interaction between cell and environment, the *exposure* of the cell to the environment and its ingenuously automatic reaction to environmental changes, and the maintenance of this automatism by means of the metabolism of nutrients. The further development of the theme, comprising the automatism in the cell system and organisms *in toto*, interacting with a variable environment, has some features of rough generalization, which, however, is intentional here with the aim of relating the metabolic manifestations of life of the individual cell to the network of action of all cells, cell systems, and organisms within the biosphere. In order to carry out this discussion, certain descriptions of cellular morphology and chemistry had to be treated very summarily. Particularly the genetic aspects are given deplorably little attention here, as well as the role of oxidative phosphorylation in the energy metabolism. Both would require space far out of proportion to the contents of the rest of the chapter.

Chapter 4 represents a breather with a presentation of a time scale and a somewhat academic discussion of appropriate methods and mental attitude for analyzing the origin of life in general. Here, delineated for the first time, is the thread along which the origin of life on our earth is reconstructed: first is the study of the formation of the metabolism of carbon compounds in early times, followed by the formation of active units within the framework of the general process as it is gradually modified over the course of billions of years.

Chapter 5 is purely a description of the environment of the Cambrian, to emphasize the fact that already a fully modern form of life-as-organisms and life-as-a-whole exists. At the same time, life in the Cambrian is identified as the outpost to which the further reconstruction of life in prebiotic times will in due course be connected. An attempt to reconstruct something of pre-Cambrian life by indirect methods, especially isotope studies, is presented as an interlude in *Chapter 6*. This entire branch of natural science, which

Rankama calls isotopic geology, is a fascinating study with many untapped possibilities. The data presented are all from Rankama's excellent monograph. With this glance toward a vaguely formed organic life in the pre-Cambrian, the following analysis is based on reconstruction of the modifications that can be expected of the early carbon material accumulated in the first formative stage of the planet.

Chapter 7 is an attempt to trace the condensation process that gave rise to our planetary system, including our earth. For those who have the suitable background, the works of Alfvén concerning this process are productive reading. Of special interest here is, naturally, why the formed planets exhibit such different chemical characteristics, which Alfvén's work may give a clue to. The import of the chapter lies in the presentation of stellar and interstellar material and the chemical possibilities related to the cooling of the hot ionized material down to temperatures between 0° C. and 2,000° C. The possibilities of cyanogen compounds after a first cooling and accumulation on a young earth and their further role as participants in a primitive conversion of carbon compounds are my own ideas, which are developed here in some detail. On the whole, the purpose of the chapter is to illustrate the possibilities of a first flux of carbon compounds between stages of extreme destruction in high heat and subsequent condensation in cooler regions, with a reconstruction of the possible composition of the material after it accumulates as the primary raw material for further conversion reactions around the surface of the earth. Perhaps it should be pointed out that the chemical reactions presented here are not at all colored by any science-fiction treatment of the material. They are extremely real transformations, elementary organic chemistry, which simply *must* have taken place to a greater or lesser extent at the stated moment, or, more correctly expressed, epoch.

Chapter 8 elaborates further on the theme of transformation of carbon compounds in a considerably cooler environment, the epoch when the earth cooled down to an average temperature not appreciably different from today's. Regarding condensation of water in the first phase of this period, there are no certain data at all, but geologists are inclined to assume that considerably less water in the form of seas existed in archaic times compared with the present. Here I have been influenced to some extent by a comprehensive article by W. W. Rubey, "Geologic History of Sea Water—an Attempt To State the Problem," in the *Bulletin of the Geological Society of America*, LXII (1951), 1111-47. The theme in this chapter is the question of the concentration of early carbon material dissolved in the seas. I am slightly skeptical of the rather optimistic theories based on chemical conversion in the seas which have been proposed by several investigators, including Oparin. The highest concentration that one can reasonably assume lies within the order of magnitude of 0.00001 per cent, a dilution that seems—at least to me—too great for any precipitations of more complex carbon material from the seas; in the manner of Oparin's coacervates. The chapter is therefore devoted to a discussion of suitable pre-concentration processes, such as the preliminary stage of a subsequent chemical conversion of the dissolved material.

Chapter 9 presents the possibilities of concentrated carbon material's being transformed into more complex molecular configurations. The theme is still purely organic chemical, without any very fanciful assumptions; data are from our present knowledge of the formation of polymers and polymeric condensates from simpler components. In my opinion, the important thing here is to suggest the geologic possibilities of localized separation of transformed material, high molecular from low molecular.

Chapter 10 has many uncertain points. We attempt to visualize the formation of two types of environment in which the first primitive polymers might have been influenced toward dehydrogenation and hydrogenation. Regarding the latter, one can assume without any great risk that local concentration of hydrogen sulphide here and there might have led to a continuous change in the composition of the polymeric material in the direction of more complex structures. The former reaction has been localized to certain regions where pronounced heat together with earlier deposited sulphur provides a chance of thermal dehydrogenation. However, we operate with assumptions concerning the formation of the archaic environment on the whole that may be dubious. Certainly there were localities of escaping hydrogen sulphide, then as well as now, and likewise there probably were volcanic regions where organic material occasionally was confronted with sulphur at moments of suitable heat. Also, some localities probably contained a certain amount of iron on the three-plus level, others on the two-plus level, which probably influenced certain organic compounds locally. The difficulty, however, is to find a realistic solution to the problem of how *both* reaction tendencies could function at the same time, giving an early polymer the characteristic of destruction by dehydrogenation and reconstruction by hydrogenation, which is peculiar to later living units. In an attempt to find a way out of this dilemma, I visualize in *Chapter 11* a local environment where both tendencies can take place. This is envisaged in the section on local water accumulations, where subtle interaction between hydrogenating and dehydrogenating reactions may conceivably take place. The idea of the tarn, with its polymeric material, as a first living unit per se is my own, and it seems to me that this idea has certain possibilities of development.

Chapters 12 and 13 advance the idea that the first development of living units goes from larger polyfunctional organizations to smaller, more specialized. The development of photosynthesis as a logical consequence of hydrogen-sulphide-influenced carbon-dioxide-fixation in earlier stages is a fairly justified assumption, likewise the scheme of the development of the stabilizers for cellular metabolism that we today call genes. At the end of *Chapter 13* we connect the discussion concerning the prebiotic history of organic material to the facts of organic life in the Cambrian. Further development falls beyond the scope of this book; it concerns paleontologic facts, which stand on considerably firmer ground than this somewhat ambitious reconstruction of life, the earliest development of which has not left behind any traces in the form of well-documented fossil remains.

The theme of this exposition of the origin of life on our earth is strictly deterministic. How could it be anything else if one adheres to the truth that lies in the nature of the material that is and develops into life? The developmental possibilities of matter are fascinating. It is possible that at times we underestimate them, particularly when we unconsciously or consciously glorify our own special development, which at times seems to contain something intentional. My own opinion is that, regardless of whether a purely material development has given us a chance to become what we have become, the result of it all gives us every reason to accept gratefully what is. If we are in the mood for it, there is so much around us to appreciate of the different forms of expression of life, both subtle and more obviously dynamic. Colors, forms, music and motion, illusions and realities. All of it.

INDEX AND GLOSSARY

[Figures in parentheses refer to page numbers.]

Adaptation (50). The ability of a cell to utilize and convert substances ordinarily found in its environment. The mechanism entails an accentuated synthesis of enzymes suited for this activity where the foreign substances to which the cell is adapted structurally influence the enzyme synthesis in a certain direction.

Adsorption (113). The ability of a high molecular substance to bind certain low molecular substances on its surface. In certain cases the adsorption can be specific for certain substances. Enzyme reactions are based on adsorption of the substances that are affected in one direction or another.

Aerobe (38). A cell which functions chemically in the presence of oxygen. The opposite of anaerobe.

ALFVÉN, H. (88, 149, 152)

α *(alpha) particle* (73)

Amino acids (30). Nitrogenous organic substances, constituting structural units of proteins. A number of amino acids linked together during the splitting-off of water build long chains, peptides, that, coiled in a special manner, form different protein structures.

AMP, ADP, ATP (36, 132). Abbreviations of adenosine-monophosphate, -diphosphate, -triphosphate. The last mentioned, with its three phosphate units in succession, is the main conveyor of chemical energy within living cells. In the metabolism of nutrients within cells of all types, the resulting energy is stored in newly formed ATP, produced from ADP and phosphate under certain conditions. ATP can later transport the bound energy to other substances in and for the formation of new chemical combinations.

158 INDEX AND GLOSSARY

Anaerobe. See *Aerobe.*
ASTON, F. W. (72, 73)
Atomic weight (72)
BERNAL, J. F. (113, 148)
β *(beta) particle* (74)
BLUM, H. F. (147)
Cambrian (60). Early epoch in the history of organic life seen from a geologic point of view. The period lies between 500 and 400 million years before our time.
Carbohydrate (60). General term for sugars, simple and complicated. In the complex carbohydrates there are different patterns of simple sugar units, e.g., glucose, linked into larger structures. Our common "sugar" contains two units in combination: glucose and fructose; see also *Cellulose.* Complicated carbohydrates are converted into simpler ones in the cell by enzymatic hydryolysis. Further decomposition leads gradually to pyruvic acid.
Carbon dioxide acceptors (16, 40). In photosynthesis an essential moment is the binding of carbon dioxide to certain substances in and for the combination's further hydrogenation and conversion to sugar. It is assumed that in the main part of the green plant cells the actual carbon dioxide acceptor is a simple sugar with 5 carbon atoms and a phosphate group attached to each end of the molecule. The technical name is ribulose diphosphate, which in Figure 11, p. 41, is called RP.
CARSON, RACHEL (103)
Catalysts (3, 26). Substances that activate a chemical reaction by participation without being altered themselves. Enzymes are to be regarded as cellular catalysts for effectuating the conversion of nutrients and the synthesis of cellular substance. In many cases we encounter in cells exceedingly complicated systems of co-operating catalysts of the enzyme type.
Cellulose (14). Essentially a linear molecular structure containing coupled units of glucose, a simple sugar. Cellulose in combination with lignin is a component of wood. Other structural variants of linked glucose residues are starch and the glycogen of the liver.
Chlorophyll (14, 134). A complicated porphyrin with a centrally located magnesium atom and a long hydrocarbon structure attached at one site in the molecule (see *Porphyrins*). Chlorophyll is a component of the organizations in green plant cells that effectuate photosynthesis, those we call chloroplasts (see *Photosynthesis*). Chlorophyll has the function of mediating the utilization of light energy in the cleavage of water into oxygen and acceptor-bound hydrogen (see *Hydrogen acceptors*).

INDEX AND GLOSSARY 159

Chromosomes. Linear organizations of proteins and nucleic acids of DNA type within living cells, most often observable within the cellular region called the cell nucleus. See also *Genes; Nucleic acids.*

CLUSIUS, F. (80)

Cytochromes (134). Certain combinations of proteins with complex ring structures, hemins, chemically related to hemoglobin of the blood. Cytochromes have a special function in cells where acceptor-bound hydrogen is confronted with oxygen resulting in the formation of water and a simultaneous gain in energy, both thermal and chemical. See *Oxidative phosphorylation.*

Decarboxylation (27). The removal of a molecule of carbon dioxide from an organic compound. This can take place purely chemically or in living cells by the action of special enzymes, decarboxylases.

Decomposition (21). Simplification of matter; opposite of *synthesis*. See *Synthesis; Photosynthesis; Respiration.*

Dehydrogenation (27, 95). The removal of hydrogen from a carbon compound. This can take place purely chemically or in living cells by the action of enzymes. Dehydrogenating enzymes are called dehydrogenases. The opposite of dehydrogenation is hydrogenation.

DNA (46). Abbreviation of deoxyribonucleic acid (see *Nucleic acids*).

Electron (71 ff.)

Element (20, 71 ff.)

Energy (4, 15, 33)

Enzymes (26). Protein substances with the special ability of binding certain substances in their molecular structure; compare *Adsorption*. In the combination of an enzyme with another substance, the linkage is most often markedly specific, when at the same time the bound substance is broken down in some way or combined with another. Most chemical conversion processes that take place in a living cell, those that lead both to the decomposition of nutrients and to the synthesis of cellular material, are effectuated by different specific enzymes.

Evolution. A term covering the long-term changes of a certain type of organism during a long series of generations. The concept "evolution" embraces an exceedingly complex interaction of factors: environmental influence, adaptive processes, different selective processes, genetic factors. An accidental alteration of the last-mentioned, mutations, can at times produce rapid changes in the special type of the organism which remain and are modified by other slowly acting factors.

Fat (27). General designation for substances containing long chain structures with linkages of $-CH_2$ type, sometimes branched, the majority water-insoluble, "fat-like." A general type of fat is a combination of 3 fatty acids with the alcohol glycerol. Variants of this theme are components of the structural elements in the brain and nerves. In the living cell, fat can be broken down by certain hydrolases and resynthesized by other enzymes (see *Hydrolysis*).

Fatty acids (27). Special hydrocarbon structures, often with long chains of -CH_2 type, straight or branched. To these structures are attached one or more -COOH groups, which give the fatty acids their acidic character. The simplest fatty acid is formic acid, HCOOH; acetic acid is CH_3COOH.

FISCHER, E. (110)

Fossil carbon (14). Layers of material from former organisms, partly of carbonate type, remains of calcium carbonate shells, partly converted organic material of complex type. Anthracite is plant material converted under pressure and heat for longer or shorter periods, probably chiefly wood. Petroleum is another form of fossil non-carbonate carbon, probably derived from isolated deposits of marine organisms exposed to evaporation and high pressure. Fossil non-carbonate carbon contains high amounts of carbon and hydrogen, comparatively little oxygen and nitrogen.

FRUTON, F. (149)

Genes (46, 138). Certain limited molecular regions within the chromosomal organization of a cell, with directive influence on the synthesis of the cell's supply of special enzymes. It is probable that highly organized DNA (see *Nucleic acids*) is the active agent here, affecting the synthesis of different forms of RNA and thereby the protein synthesis in general. In many cell types it has been possible to localize the position of the genes within the chromosomal groups, which most often are found in the cell nucleus. A mutation is a particular occurrence within a region of chromosomal material resulting in alteration or destruction of a gene, which has long-lasting consequences for the chemical system of the cell during subsequent generations.

GOLDSCHMIDT, V. M. (149)

HALDANE, J. B. S. (139)

Half-life (73). The time required for a radioactive element to be reduced by disintegration to one-half its original weight.

Hemins (134). Certain complex 4-ring systems of porphyrin type with an iron atom symmetrically linked to the 4 nitrogen atoms (see also *Porphyrins*).

HEVESY, G. VON (80)

Homeostasis (51). A term covering the ability of cells and organisms to maintain relatively uniform conditions within themselves in spite of changes in the internal and external environments. This exceedingly complex mechanism can find expression in rapid responses, e.g., all our reflex reactions. Among the slower reactions of homeostatic nature, adaptation is often the sustaining factor.

INDEX AND GLOSSARY 161

Hydrides (89, 93). General term designating simple substances containing hydrogen in combination with another atom, most often a metal.

Hydrogen acceptors (16). The substances that in a cell incorporate the hydrogen split off in enzymatic dehydrogenation reactions. Several types exist; some are related to the vitamin nicotinamide, others are related to the vitamin riboflavin. The former are generally the primary hydrogen acceptors, which often transfer the bound hydrogen to flavins, which in their turn are confronted with cytochromes.

Hydrogenation (28, 125). Incorporation of hydrogen in an organic substance. Hydrogenation can take place under chemical conditions, often under the influence of certain catalysts. In the living cell, a given hydrogenation is often managed by the same enzyme that effectuates the corresponding dehydrogenation.

Hydrolysis (28). The action of water on a substance, whereby the substance during the incorporation of water is changed in one direction or another, most often with the splitting-off of some radical. Hydrolysis of proteins involves the cleavage of their peptide groups into amino acids by stepwise incorporation of water. Fat can be hydrolyzed to glycerol and fatty acids, complex carbohydrates to simpler types of sugar. Within a cell, hydrolysis functions under the influence of special enzymes, hydrolases.

Iron bacteria (127). Primitive micro-organisms capable of utilizing salts of iron of different oxidative levels as the source of energy for growth.

Isotopes (72). Different variants of one and the same atom, chemically extremely alike, but with varying atomic weight.

KUIPER, G. P. (87, 149)

Labile phosphates (34, 36). See *AMP, ADP, ATP*. Among the variants that participate in living cells in the phosphate metabolism and thereby in the energy metabolism as a whole, there are at least eight types related to ADP and ATP, plus others that are to be regarded as systems for the storage of chemical energy, e.g., creatine phosphate. Also metaphosphates are included under the term "labile phosphates" because of their property of being hydrolyzed into simple, stable phosphate.

LIBBY, W. F. (76)

Lignin (14). A very resistant component of wood combined with cellulose. Like cellulose, the lignin molecules are a chain pattern of many units. These contain in general a 6-ring structure of hydrocarbon nature, plus an attached chain containing 3 carbon atoms in a row. There is a great number of variants of this theme.

Mass (72)

Metabolism (24 ff.). The decomposition of nutrients within a cell and the conversion of the products thereof into cellular material. Also, the decomposition of the latter into simpler material within the dynamic action of the cell is included in the concept "metabolism."

Metaphosphates (34 ff., 109). See *Labile phosphates*.

MILLER, S. L. (96)

Mitochondria (37, 47). Complicated structures containing organizations of enzymes that in cells, particularly those living in an aerobic, oxygenous environment, effectuate the total decomposition of certain substances into carbon dioxide and water. Participating enzymes within mitochondria in general are partly dehydrogenating enzymes, partly decarboxylating. Further, there are enzymes that mediate the contact of the split-off hydrogen with oxygen to form water. Cytochromes are among these (see *Cytochromes*).

Mutations (46). See *Genes*.

Neutron (71 ff.). Uncharged component of atomic nuclei (see text).

NIER, A. O. (78, 80)

Nucleic acids (45). Structure and function (see text).

OPARIN, A. I. (102, 109, 147, 153)

Organelles (123 ff.)

Organic material (17). Product of photosynthesis or carbon dioxide fixation.

Organisms (123 ff., 128)

Oxidation (37, 95). In its strictest sense, the concept "oxidation" embraces chemical reactions where one or more atoms of oxygen are incorporated into a molecule. In a living cell we have many examples of such reactions, directed by enzymes. The main part of a nutrient that is decomposed within a cell, however, is not oxidized directly; it is dehydrogenated by the splitting-off of hydrogen. The acceptor-bound hydrogen is then oxidized by oxygen to water. The carbon dioxide that is given off from a living cell is thus not a product of oxidation in the real connotation; it is the result of a splitting-off of carbon dioxide from dehydrogenated products (see p. 27).

Oxidative phosphorylation (37). The process that in mitochondria converts acceptor-bound hydrogen into water via the confrontation with oxygen, coupled with formation of labile phosphates, chiefly ATP, for further activation of innumerable processes within the cell (see *ATP*).

Oxides (95). The term most frequently applied to simpler inorganic substances with one or more atoms of bound oxygen.

Peptides (43, 111). Long chain structures of amino acid units (see *Amino acids*). The structures are components of proteins of all kinds.

Photosynthesis (15 ff.). Term designating the ability of green plants under the influence of light to incorporate carbon dioxide and water for the synthesis of cellular material. This process, photosynthesis, or, in everyday language, carbon dioxide fixation, takes place in specially

organized structures in green plants—chloroplasts containing chlorophyll (see *Chlorophyll*).
Phthalocyanines (118). Highly stable substances probably formed in archaic times, closely related to hemins. See *Hemins*.
Phytoplankton (64). Comprehensive term for marine plants, those of small dimensions with the ability of photosynthesis (see *Photosynthesis*).
Polycondensates (108). Long molecular structures with similar but somewhat varying units within the structure. Proteins are polycondensates.
Polymers (107, 114 ff.). Long molecular structures with one and the same unit repeated within the structure.
Porphyrins (133). Symmetric molecular structures of 4 rings, each containing one nitrogen atom and 4 carbon atoms. In addition, the arrangement contains a linking carbon atom between the rings and a number of side groups of varying nature. Often a porphyrin can occur with a metal atom symmetrically bound between the nitrogen atoms. Porphyrins constitute fundamental structures in cytochromes, hemins, and chlorophyll (see *Cytochromes; Hemins; Chlorophyll*).
Pre-Cambrian (83). The epoch in the history of the earth prior to 500 million years before our time. The beginning of this period is somewhat uncertain; however, the period can be considered to have lasted from 1,500 to 500 million years ago. Very few remains of organic life exist from this time.
Proteins (44). Very complicated structures based on specific coiling of long peptide chains composed of amino acid units (see *Amino acids; Enzymes; Peptides; Polycondensates*). All enzymes are proteins. The constituent amino acid units can vary between fifty and many hundreds. The variability in proteins underlies the variability of structure and function that we can observe in all living cells and organisms.
Proton (71 ff.). Positively charged component of atomic nuclei (see text).
Pyruvic acid (28). Intermediary product in carbohydrate decomposition, the key substance in cellular metabolism.
Radioactivity (73 ff). Atomic decay caused by internal instability.
RANKAMA, K. (78, 79, 80, 81, 82, 83, 149, 152)
Reduction (114). Frequently used in the same connotation as hydrogenation. A reducing atmosphere implies that hydrogen and highly hydrogenated substances dominate. Reduction, in the sense of hydrogen incorporation, is often used as the opposite of oxidation in the sense of oxygen incorporation.
Respiration (15). Release of carbon dioxide and water; opposite of photosynthesis.
RNA (45, 139). Abbreviation of ribonucleic acid (see *Nucleic acids*).
RUBEY, W. W. (153)
RUSH, A. L. (147)
SAHAMA, T. G. (149)

SEDERHOLM, J. J. (82, 83)
SIMMONDS, S. (149)
Specificity (26). Term frequently used to emphasize the selective action of enzymes on the substances they affect (see *Enzymes*).
Substrate. The same as nutrients in general; specifically, a substance upon which an enzyme acts.
Sulphur bacteria (125). Micro-organisms with the ability of utilizing sulphur compounds of different kinds, at times hydrogen sulphide, as the source of energy for their metabolism.
Synthesis (17). Complication of matter. See *Photosynthesis*.
Thermal dehydrogenation (95). Expression characterizing certain chemical reactions, in which a carbon compound is made to split off hydrogen by means of heat treatment.
Thermonuclear (87). Term for such processes during which, under exceedingly high temperature, certain atomic nuclei can be made to coalesce —fuse—with the liberation of very great amounts of energy. The detonation of a hydrogen bomb is the result of a thermonuclear reaction. In all suns the fusion of hydrogen atoms to helium is the fundamental principle for the generation of energy.
THODE, H. G. (82, 83)
UREY, H. C. (80, 84, 149)
WEIZSÄCKER, C. F. VON (87)
WICKMAN, F. (78, 80)

PRINTED IN U.S.A.